Synthese Library

Studies in Epistemology, Logic, Methodology, and Philosophy of Science

Volume 504

The aim of *Synthese Library* is to provide a forum for the best current work in the methodology and philosophy of science and in epistemology, all broadly understood. A wide variety of different approaches have traditionally been represented in the Library, and every effort is made to maintain this variety, not for its own sake, but because we believe that there are many fruitful and illuminating approaches to the philosophy of science and related disciplines.

Special attention is paid to methodological studies which illustrate the interplay of empirical and philosophical viewpoints and to contributions to the formal (logical, set-theoretical, mathematical, information-theoretical, decision-theoretical, etc.) methodology of empirical sciences. Likewise, the applications of logical methods to epistemology as well as philosophically and methodologically relevant studies in logic are strongly encouraged. The emphasis on logic will be tempered by interest in the psychological, historical, and sociological aspects of science. In addition to monographs *Synthese Library* publishes thematically unified anthologies and edited volumes with a well-defined topical focus inside the aim and scope of the book series. The contributions in the volumes are expected to be focused and structurally organized in accordance with the central theme(s), and should be tied together by an extensive editorial introduction or set of introductions if the volume is divided into parts. An extensive bibliography and index are mandatory.

Emiliano Ippoliti • Melissa Vergara-Fernández •
Fabrizio Zennaro
Editors

Philosophy and Finance: Ten Open Questions

Springer

Editors
Emiliano Ippoliti
Department of Philosophy
Sapienza University of Rome
Rome, Italy

Melissa Vergara-Fernández
University of Groningen
Groningen, The Netherlands

Fabrizio Zennaro
Banca d'Italia
Rome, Italy

ISSN 0166-6991 ISSN 2542-8292 (electronic)
Synthese Library
ISBN 978-3-031-96509-8 ISBN 978-3-031-96510-4 (eBook)
https://doi.org/10.1007/978-3-031-96510-4

This Springer imprint is published by the registered company Springer Nature Switzerland AG
The registered company address is: Gewerbestrasse 11, 6330 Cham, Switzerland

Contents

Chapter 1
Philosophy and Finance: Ten Open Questions

Emiliano Ippoliti, Fabrizio Zennaro, and Melissa Vergara-Fernández

Abstract This introductory chapter briefly outlines key elements of the relationship between philosophy and finance. The discussion then centers on several of the main areas where philosophy contributes to finance, highlighting a few relevant issues in ethics, ontology, methodology, technology, and politics. Finally, an overview of the trends and prospects of the philosophy of finance that emerge by the contributions to this volume is presented.

1.1 Finance and Philosophy: Some Central Themes

Financial markets are a central element of modern economies. They determine the allocation of economic resources not only across space, but also across time and thus, per definition, in an uncertain environment. Their influence on society is therefore profound. The rise of passive investing, for instance, has arguably democratized investment by fostering the possibility for individual households to participate in these markets. Or, as John Kay has recently put it, "to see a modern capitalist, perhaps you should look in a mirror in the home you own. Or take a selfie" (Kay, 2024). On the downside, the Global Financial Crisis also demonstrated the terrible negative effects they can have. Philosophers, and the emerging field of philosophy of finance (see e.g. De Bruin et al., 2020; Ippoliti, 2020b; Sandberg & Warenski, 2023), which is the central focus of the book, play a valuable role addressing the challenges that these markets pose by providing frameworks for

E. Ippoliti
Department of Philosophy, Sapienza University of Rome, Rome, Italy

F. Zennaro (✉)
Banca d'Italia, Rome, Italy

Sapienza University, Rome, Italy
e-mail: fabrizio.zennaro@uniroma1.it

M. Vergara-Fernández
University of Groningen, Groningen, Netherlands

E. Ippoliti et al. (eds.), *Philosophy and Finance: Ten Open Questions*, Synthese Library 504, https://doi.org/10.1007/978-3-031-96510-4_1

understanding both financial markets and their ontology and ethics, as well as the methodology, and epistemology of finance, the science which studies financial markets (see for example Endörfer, 2022; Ippoliti, 2023).

These issues are intertwined, as the chapters in the volume show: for example, ethical issues involve technological concerns, which rise ontological problems, and vice versa. Take money: what is money (an ontological issue, see Guala's chapter in this volume) is tied to the technological innovation (as Zennaro's chapter in this volume shows), and to the ethical issues it raises, especially those regarding financial inclusion and financial divide.

To illustrate the value of philosophy in addressing these issues, let's examine a few of them in more detail.

1.1.1 Some Ethical Issues in Finance

One of the most important areas where philosophy contribute to finance is ethics (see De Bruin et al., 2020 for an overview). Finance is a complex and dynamic field, and there are many ethical problems that arise during financial transactions, settings, and designs.[1] Philosophers can help to clarify these problems and develop ethical frameworks for the financial industry. Philosophers can help on key ethical issues in finance, as *deception and fraud,* by developing standards for ethical behavior in the financial industry; strengthening guidelines for resolving conflicts of interest and ensure that financial professionals act in the best interests of their clients; defining what constitutes insider trading and develop penalties for those who engage in it (see e.g. Irvine, 1987)*; or*, identifying the elements that lead to preferring short-term strategies over long-term ones.

Moreover, philosophy can help dealing with issues emerging from regulation in finance, such as the appropriate role of government regulation in financial markets, or how regulation can be designed to promote stability and fairness.

Some of these issues are deeply intertwined with technological issues (see § 1.4.), as the chapters in this volume show. One example is digital divide and financial inclusion. As finance becomes increasingly digitized, there are philosophical debates about financial inclusion and the potential exclusion of certain demographics due to technological barriers or disparities in access to financial services served by innovative technology. Another example is the ethical use of financial technology and fintech advancements. This includes ethical implications surrounding robot-advisors, automated decision-making, biases in algorithmic systems, and the responsibilities of financial institutions in ensuring fairness and transparency.

[1] On this point see Angel & McCabe, 2009; James, 2012; Mejia, 2022; Sandberg, 2008,

1.1.2 Some Ontological Issues in Finance

A key questions that the philosophy of finance grapples with is the ontology of finance, which is the study of the nature of financial objects and concepts. There are numerous open questions in the ontology of finance (see e.g. Sansò & Smith, 2023; Varzi, 2008), concerning the fundamental nature of entities like money, risk, value, rationality, agency, models[2] and efficiency within financial systems. For instance, there is debate about the nature of money itself: How is money related to value, or what is the nature of financial assets, their differentiation from other assets, and the classification of various asset types like stocks, bonds, and derivatives? The concept of risk, its definition, measurement, and categorization into types like market, credit, and operational risk, is another area of ontological interest. The creation and measurement of value in finance, as well as the relationship between price and value, are further ontological questions. The problem of agency, particularly the mitigation of conflicts between managers and shareholders, is another area of ontological exploration. Finally, the nature of bubbles and crashes, their formation, causes, and consequences, are topics of significant ontological interest (see e.g. Ippoliti, 2023, 2024b).

As the philosophy of finance continues to develop, it is likely to shed further light on the ontological issues that underpin the field of finance. This will help us to better understand the complex and often paradoxical nature of financial markets and its institutions.

1.1.3 Some Methodological Issues in Finance

Philosophy of finance can also help to address the methodological challenges posed by finance, and the role of mathematics in shaping it.[3] Finance is a data-intensive field, and there is a growing debate about the role of models (see e.g. Morgan, 2012; Boldyrev & Ushakov, 2016; Vergara-Fernández et al., 2023a) and data in financial analysis (see e.g. Ippoliti, 2019). Philosophers can help to clarify the limitations of models and develop more rigorous methodologies for financial analysis, including the problem of predictions[4] and its relation to the performativity[5] of the tools employed to generate these predictions (see MacKenzie, 2006; Vergara-Fernández et al., 2023b).

For example, we have at least two recent key methodological issues in finance. The first one is the role of models: financial models are often used to predict future

[2] See e.g. Van der Heide, 2019.

[3] On this point see Ippoliti, 2020a.

[4] On this point see also Ippoliti, 2024.

[5] On performativity and finance see Callon, 1998; MacKenzie, 2003, 2006, 2017; MacKenzie & Millo, 2003; MacKenzie & Spears, 2014; Marti & Gond, 2018; Svetlova, 2012; Walter, 2016.

market movements, but these models can be unreliable or interfere with the financial processes they aim at describing. Philosophers can help to assess the limitations of models and develop more rigorous methods for model validation. The second is the role of data in financial analysis: financial markets are constantly generating new data, and it is important to develop rigorous methods for analyzing this data. Philosophers can help to improve the quality of financial data and develop new tools for data analysis.

1.1.4 Some Technological Issues in Finance

The intersection of technology and finance within the realm of the philosophy of finance has highlighted several crucial issues (see e.g. Pardo-Guerra, 2019). The rise of algorithmic trading and AI has raised questions about the moral and ethical implications of using these tools in financial decision-making, particularly regarding their impact on market stability and the potential for market manipulation. Additionally, the integration of big data analytics in finance has brought to light concerns about privacy, data security, and the ethical use of personal information in predictive modeling and risk assessment. High-frequency trading, with its rapid execution of trades based on algorithms and computational power, has raised questions about its impact on market volatility, fairness, and stability. The emergence of blockchain technology and crypto-assets has posed challenges and opportunities in finance, prompting philosophical inquiries about decentralization, transparency, regulation, and the role of digital currencies in the financial system. Moreover, the philosophy of finance delves into the ethical implications of fintech advancements, including the use of robo-advisors, automated decision-making, and biases in algorithmic systems. The rapid pace of technological innovation in finance often outpaces regulatory frameworks, leading to philosophical discussions about the need for updated regulations, their adaptability to technological advancements, and the balance between fostering innovation and ensuring financial stability. Finally, as finance becomes increasingly digitized, there are philosophical debates about financial inclusion and the potential exclusion of certain demographics due to technological barriers or disparities in access to financial technology.

These issues underscore the complexities and ethical considerations arising from the integration of technology into finance. Philosophical inquiries in the field aim to analyze, critique, and propose ethical frameworks that navigate these challenges while fostering a fair, transparent, and stable financial system.

1.1.5 Some Political Philosophy Issues in Finance

Financial markets are also a major influence on the political sphere. The rise of financialization has led to a growing concentration of wealth and power in the hands of a few, which has raised concerns about political influence. Philosophers can help to analyze the political implications of finance and develop policies and institutions (see e.g. Guala, 2016) to address these concerns (see e.g. Dietsch, 2010; Herzog, 2017a, 2017b).

We have at least two key political issues in finance. First, the power of financial markets. Since financial markets have a significant impact on the political process, and large financial institutions can wield considerable influence over governments, philosophers can help to examine the relationship between finance and political power and develop policies to ensure that financial markets are not used to undermine democracy. Second, the role of finance in global governance: financial markets are increasingly globalized, and this has raised concerns about the ability of national governments to regulate them. Philosophers can help to develop international cooperation mechanisms to address the challenges posed by globalized finance.

1.2 The Future of Finance and Philosophy: The Ten Open Questions

The role of philosophy in finance is likely to become even more important in the future.[6] As finance becomes more complex and data-driven, the need for rigorous frameworks for understanding the mentioned issues raised by finance will only grow.

This book brings together diverse analyses addressing key questions and challenges in the financial sector, illustrating how philosophy can contribute to finance and vice versa. The collected chapters explore issues poised to shape the future of finance and, consequently, our economic and social landscapes, given the growing dependence of global economies on financial systems. The book examines significant topics such as the connection between financial systems and climate change, essential financial ontologies for market design, the role and impact of AI in finance, the nature and potential of money, including digital currency, methodological challenges in applying models and mathematics, and critical assessments of core financial theories like the efficient market hypothesis. As such, the book is relevant to philosophers specializing in finance, science, and economics, as well as to, practitioners, and graduate students interested in the methodological, ontological, epistemological, and ethical dimensions of finance. It provides thought-provoking discussions that reveal the limitations of conventional perspectives on knowledge production, modeling, and ethics in financial markets, while emphasizing the importance of philosophy in the design and operation of these markets.

We outline ten key themes relevant to this context, providing background and discussing how the included chapters contribute insights into each (§1.3).

[6] See e.g. Ferguson, 2008.

The first theme addresses how an enhanced understanding of financial ontology can improve data management within the financial system (see Sansò's work).

The second theme explores the influence of AI on financial markets, particularly how information dissemination aligns with the Efficient Market Hypothesis (see Barucca & Morone). The third focuses on empirical studies concerning the nature, perception, and use of money (see Noyes, Guala & Keil).

This is linked to the evolving concept of digital money, highlighting that it is more than just a technological advance (see Zennaro).

The fifth theme examines the epistemic character of prices and quotes, emphasizing that market design is inherently theory-driven and not neutral (see Ippoliti).

The sixth theme considers the nature of financial models, specifically the concept of the 'ontology of models' and insights from models used in Libor derivatives markets (see Spears).

The seventh theme, of a methodological nature, discusses how agent-based modeling could encourage a paradigm shift in finance and economics (see Fabretti).

Another theme focuses on market design, addressing the benefits, drawbacks, and methodological considerations of trading on alternative venues like 'dark pools' and their impact on standard markets (see Altay, Colaneri & Eksi-Altay).

This is followed by a discussion on algorithmic trading, its various forms, and the associated pros, cons, and methodological challenges (see Cellai & Puorro). Finally, the volume considers the structure and debates surrounding mainstream financial theory, particularly the Efficient Market Hypothesis (see Walter).

The contributions to this volume discuss the problems, drawbacks, and prospects of these issues, and provide suggestions for rethinking the financial system and the theories developed to characterize it.

1.3 An Overview of the Chapters

This volume brings together scholars from various academic and professional backgrounds to examine philosophical questions in finance, such as the nature of money, the function of markets, and the ethical dimensions of financial practices. The chapters offer a wide array of viewpoints and contribute meaningfully to the expanding field of the philosophy of finance. A prominent theme throughout is the evolving nature of finance, driven by developments like fintech, data analytics, and the globalization of markets. These transformations prompt important philosophical questions about finance's role and impact on society. Another significant theme is the intersection of finance and ethics, emphasizing the importance of considering the moral consequences of financial actions due to their profound effects on people's lives. Collectively, these chapters provide new understandings into central issues within the philosophy of finance and pose critical questions regarding its future direction.

In particular, in their chapter ***How low-cost AI universal approximators reshape market efficiency***, Paolo Barucca and Flaviano Morone question the efficient market hypothesis (EMH), which famously stated that prices fully reflect the information available to traders. Since this critically depends on the transfer of information into prices through trading strategies, traders optimize their strategy with models of increasing complexity that identify the relationship between information and profitable trades more and more accurately. Barucca and Morone argue that, under specific conditions, the increased availability of low-cost universal approximators, such as AI systems, should be naturally pushing towards more advanced trading strategies, potentially making it harder and harder for inefficient traders to profit. In particular, they leverage on a generalized notion of market efficiency, based on the definition of an equilibrium price process, that allows them to distinguish different levels of model complexity through investors' beliefs and trading strategies optimization, and to discuss the relationship between AI-powered trading and the time-evolution of market efficiency. Thus, they outline the need to describe out-of-equilibrium market dynamics in an adaptive multi-agent environment.

In their chapter ***A tour into dark pools,*** Sühan Altay, Katia Colaneri, and Zehra Eksi-Altay examine the structure and function of dark pools—private trading venues that offer confidentiality and anonymity to participants, which primarily include institutional investors. These pools execute trades away from the public eye, preventing the disclosure of large order details before execution and thus minimizing adverse price effects, such as the market price impact seen in standard lit markets. Unlike public exchanges like the NYSE or Nasdaq, dark pools do not show pre-trade order details, maintaining secrecy until trades are completed. This anonymity helps prevent predatory trading practices that can negatively impact prices.

They argue that despite the advantages, including better price execution for large trades, dark pools raise concerns related to transparency, fairness, and conflicts of interest. Regulatory bodies like the SEC, FINRA, and ESMA have imposed rules to address these issues. Altay, Colaneri, and Eksi-Altay also highlight the evolution of dark pools, including the emergence of second-generation lit-dark pools managed by major investment banks, and examine the ethical and regulatory implications associated with these trading environments.

In her chapter ***A critical discussion on how agent-based modeling participates in a potential paradigm shift in Economics***, Annalisa Fabretti discusses the historical evolution of economic theory and the emerging shift towards a more complex and realistic modeling approach in economics. Traditional economic modeling, rooted in the works of Walras, Pareto, and later Arrow, Samuelson, and Lucas, is characterized by analytically tractable models featuring perfectly rational, representative agents. These models simplify economic behavior by assuming homogeneity among agents, which has long been the dominant paradigm despite inconsistent empirical support. Recent decades have seen challenges to this mainstream approach due to advances in technology, availability of big data, and the rise of behavioral economics and econophysics. Agent-Based Modeling (ABM) has gained prominence as it allows for the simulation of heterogeneous agents with bounded rationality, capturing more realistic human behavior and complex interactions within economic systems. Unlike the representative agent framework, ABM integrates the diversity and interconnectedness of agents, linking micro-level behaviors to macro-level outcomes.

The chapter traces the mathematization of economics from the nineteenth century and explores the intersection of economics with physics, highlighting how these influences shaped the current paradigm. It then addresses the limitations of the neoclassical model, such as its inability to account for empirical inconsistencies and economic crises. Fabretti emphasizes that ABM fills these gaps by incorporating realistic assumptions, suggesting that its growing use could contribute to a significant shift in economic theory towards a more empirically grounded and nuanced understanding of complex systems. The chapter concludes by evaluating whether these developments indicate an impending paradigm shift in economics.

In their chapter ***Money on My Mind: An empirical investigation of money concepts,*** Francesco Guala, Alexander Noyes and Frank C. Keil, explore how individuals perceive and categorize money, particularly examining the distinction between current, defunct, and ad hoc (e.g., shells, cigarettes) forms of currency. Through four studies involving 1177 U.S.-based participants on Amazon Mechanical Turk, the research aims to understand whether money is seen more as a social construct or as an artifact with intrinsic characteristics. Their main key findings are at least three. The first one is '*money as a social position*': participants largely categorize ad hoc items as money if they occupied the role of currency within a societal context, showing that the concept of money relies on social functionality (e.g., being a medium of exchange) rather than on the object's physical attributes. The second is '*defunct currency*': the categorization of defunct currencies (e.g., outdated coins) is mixed, with participants showing a bimodal response. This indicates that, while some recognize such objects as no longer being money due to their lack of social currency, others consider relevant their original intent as money. The third one is '*causal reasoning*': the study concludes that people use dual reasoning, as current social practices inform the status of an object as money, while its material properties and original purpose contribute to the conceptualization of defunct currency.

The chapter highlights that money, unlike standard artifacts, is fundamentally defined by its societal role rather than its physical structure, illustrating the broader way institutions are represented cognitively.

In his chapter ***The epistemic value of a millisecond: quotes, speed bumps, and market design,*** Emiliano Ippoliti investigates the philosophical consequences of "micro-quotes," which involve placing and subsequently cancelling large orders at extremely high speeds—within milliseconds—before execution, often distributed across multiple exchanges. The chapter explores how these micro-quotes can erode the epistemic value of quotes and challenge the essential functions of financial systems. It introduces the concept of the epistemic value of a quote, defined as an agent's belief about an asset's "correct" price, noting that this "correct" price can vary based on different price theories (e.g., an objective "real" price versus a mutually agreed-upon price).

Ippoliti emphasizes that decisions regarding the target and timing thresholds for speed bumps, which are proposed solutions to issues raised by micro-quotes, are not purely technical. Instead, these choices carry significant implications for the structure and participants of financial markets, embedding epistemic and moral considerations. The chapter further delves into philosophical questions surrounding micro-quotes: what constitutes a quote when micro-quotes are allowed? Do micro-quotes diminish the epistemic value of quotes to nearly zero? Do they generate epistemic uncertainty or even lead to epistemic injustice?

In her chapter ***Why we need an ontology of finance (and how philosophers can help building it),*** Gloria Sansò argues for the development of a well-designed ontology of finance to enhance data management and transparency. It suggests that this ontology should be realism-based and aligned with the Basic Formal Ontology (BFO), although this approach requires addressing complex questions about the nature of financial entities. Philosophical inquiry, particularly around concepts like money, can inform these foundational reflections.

A significant issue revealed by the crisis was that financial institutions' IT systems were insufficient for comprehensive risk data aggregation and reporting. These failures were compounded by the lack of interoperability between institutions, which impaired the tracking of complex financial instruments like collateralized debt obligations (CDOs). As a result, when institutions such as Lehman Brothers collapsed, the market had little knowledge of how many toxic assets were held or by whom. The chapter highlights semantic technologies as a promising approach to improve IT systems, with a focus on ontologies that structure and clarify data. By building a realism-based financial ontology, the chapter posits that financial institutions could achieve better data integration and transparency, aiding in regulatory compliance and risk management.

In his chapter ***What does it mean to speak of a model's ontology? Examining the changing ontology of term structure models used in the Libor derivatives markets,*** Taylor Spears contributes to current sociological and philosophical discussions on financial modeling, particularly regarding how historical developments in financial modeling intersect with the macro-political process of financialization. Scholars have increasingly recognized that models play a significant role in reinforcing financialization by shaping practices in different fields. For instance, Chiapello and Walter (2016) highlighted how models from financial economics promoted the spread of financialized practices, while Van Der Heide (2019) demonstrated that the transfer of models from derivatives markets to the insurance sector influenced insurers' liability management.

Building on Science and Technology Studies (STS) work on ontological politics, this research underlines the importance of a model's 'ontology' in fostering financialized practices. Here, a model's ontology acts as an enduring 'core' that maintains enough stability to influence practices across various domains while being adaptable to local conditions. Despite these insights, the concept of a model's ontology remains theoretically underdeveloped. This chapter addresses that gap by examining how the practical ontology of interest rate term structure models evolved as they were adopted in OTC derivatives markets during the 1980s and 1990s. The analysis reveals that the practical ontology of these models comprises a network of assumptions, conventions, and market prices. Different components of this web have remained stable at various times, supporting the uptake of financialized practices, but the web's true strength lies in its capacity to adapt to different institutional settings from those in which the models were first created.

In their chapter ***Advanced trading systems: aggressive and unfair speculation or instruments featuring a positive impact on overall markets quality? A Comprehensive assessment*** Gioia Cellai and Alfonso Puorro share their insights on modern trading systems, commonly referred to as algorithmic traders (ATs) or high-frequency traders (HFTs). At first glance, these systems may appear unfair and detrimental, as their activities are often viewed as purely speculative and reliant on an informational and speed advantage over other, typically slower and less informed, market participants. However, ATs and HFTs are an inevitable outcome of technological advancements and form an essential part of contemporary financial markets. Cellai and Puorro argue that Advanced trading systems, such as high-frequency trading (HFT), have complex effects on financial markets. Their contributions to market quality vary with market conditions: they promote efficiency and stability during calm periods but can amplify stress during times of crisis. HFT behavior can shift rapidly from providing liquidity to aggressive profit-seeking, posing risks to other market participants and potentially leading to systemic financial events. This shift often results in wealth transfers favoring technologically superior traders at the expense of less advanced or smaller participants. Their chapter also highlights the competitive nature of the industry, where continuous innovation is crucial due to low barriers to entry and the transient dominance of firms. HFT firms invest heavily in maintaining their technological edge, driven by sophisticated human expertise. Ultimately, while modern trading systems have disrupted traditional market structures, they have also spurred adaptation among participants who seek to benefit from these advancements while managing associated risks.

In his chapter ***Market Efficiency, Risk Neutral Pricingand Choice Among Representations: a "mini-model",*** Christian Walter presents a simplified model of the Efficient Market Hypothesis (EMH) designed to clarify its main representations without requiring complex mathematics. The purpose of this model is to highlight the relationship between two key mathematical frameworks of EMH: the mean-variance model by Markowitz (1952), which operates under real-world probability (P), and the martingale pricing model by Harrison and Kreps (1979) and Harrison and Pliska (1981), which uses risk-neutral probability (Q). These frameworks have been identified by Chiapello and Walter (2016) as distinct quantification conventions.

The simplified EMH model provides an integrated understanding of these two conventions and connects them to general equilibrium theory, offering three main insights. First, it helps financial practitioners and those without specialized mathematical knowledge grasp the epistemic challenges of the EMH, enhancing its educational value. Second, it addresses one of the most conceptually difficult aspects of EMH: the uniqueness of the risk-neutral probability (Q) in a complete, arbitrage-free market. Third, it demystifies the Radon-Nikodyn derivative (L) and its role in linking the P-world and Q-world frameworks used in financial valuation. The chapter shows how this model sheds light on the prevalence of risk-neutral pricing in finance, emphasizing the conceptual issues involved in transitioning from real-world to risk-neutral perspectives. Additionally, it opens the door to exploring the ethical considerations associated with the widespread use of risk-neutral valuation methods.

In his chapter ***The new form of digital money: not a mere technological innovation***, Fabrizio Zennaro examines the intricate evolution of digital currencies, exploring the interplay of technological innovation, market forces and regulatory frameworks and argues that the future of digital currencies lies in more refined forms, including stablecoins and central bank digital currencies (CBDCs).

To argue that future forms of digital money are not merely technological innovation, Zennaro starts from an analysis of the current monetary system, identifying its peculiar characteristics. In this way, he distinguishes technological innovation between public and private currencies, arriving at the argument that other factors become relevant in the innovation process, such as competitive forces and the regulatory process. In this perspective, the chapter examines the role of 'platformization' in shaping the future of finance, where digital platforms are increasingly bypassing traditional banking intermediaries. This shift has implications for the design and implementation of digital currencies, as well as the regulatory landscape. A key question addressed is the role of public and private actors in the digital currency ecosystem. While private sector innovation can drive rapid development, public institutions, particularly central banks, have a crucial role to play in ensuring monetary policy, financial stability and resilient payment systems. The introduction of CBDCs could significantly alter the financial landscape, potentially impacting monetary policy, financial intermediation, and systemic risk. The chapter concludes by highlighting the importance of a balanced approach, combining technological innovation with sound economic principles and prudent regulation. By understanding the interplay of these factors, we can navigate the complexities of the digital currency revolution and shape a future that maximizes its benefits while mitigating potential risks.

References

Angel, J. J., & McCabe, D. M. (2009). The ethics of speculation. *Journal of Business Ethics, 90*(supp. 3), 277–286.

Boldyrev, I., & Ushakov, A. (2016). Adjusting the model to adjust the world: Constructive Mechanisms in Postwar General Equilibrium Theory. *Journal of Economic Methodology, 23*(1), 38–56.

Callon, M. (1998). What does it mean to say that economics is performative? In D. MacKenzie, F. Muniesa, & L. Siu (Eds.), *Do economists make markets?* (pp. 310–357). Princeton University Press.

Chiapello, E., & Walter, C. (2016). The three ages of financial quantification: A conventionalist approach to the financiers' metrology. *Historical Social Research, 41*(2), 155–177.

De Bruin, B., Herzog, L., O'Neill, M., & Sandberg, J. (2020). Philosophy of money and finance. In E. N. Zalta (Ed.), *The Stanford Encyclopedia of Philosophy*. (Winter 2020 *Edition*), https://plato.stanford.edu/archives/win2020/entries/money-finance/

Dietsch, P. (2010). The market, competition, and equality. *Politics, Philosophy & Economics, 9*(2), 213–244.
Ferguson, N. (2008). *The ascent of money: A financial history of the world*. Penguin.
Guala, F. (2016). *Understanding institutions: The science and philosophy of living together*. Princeton University Press.
Harrison, M. J., & Kreps, D. M. (1979). Martingales and arbitrage in multiperiod Secutities markets. *Journal of Economic Theory, 20*(3), 381–408.
Harrison, M. J., & Pliska, S. R. (1981). Martingales and stochastic integrals in the theory of continuous trading. *Stochastic Processes and Applications, 11*(3), 215–260.
Endörfer, R. (2022). *Weapons of mass destruction: Financial crises from a philosophical perspective*. Acta Universitatis Gothoburgensis.
Herzog, L. (2017a). What could be wrong with a mortgage? Private debt markets from a perspective of structural injustice. *Journal of Political Philosophy, 25*(4), 411–434. https://doi.org/10.1111/jopp.12107
Herzog, L. (Ed.). (2017b). *Just financial markets? Finance in a just society*. Oxford University Press.
Ippoliti, E. (2019). Models and data in finance: *Les liaisons Dangereuses*. In Á. Nepomuceno Fernández, L. Magnani, F. J. Salguero-Lamillar, C. Barés-Gómez, & M. Fontaine (Eds.), *Model-based reasoning in science and technology* (pp. 393–406). Springer. https://doi.org/10.1007/978-3-030-32722-4_22
Ippoliti, E. (2020a). Mathematics and finance. Some philosophical remarks, *Topoi*, pp. 1-8 (Anvur A) https://doi.org/10.1007/s11245-020-09706-1.
Ippoliti, E. (2020b). *Un Filosofo a wall street*. Egea.
Ippoliti, E. (2023). What is a financial crash? Some remarks on the ontological and epistemic preconditions. *Rivista di Estetica*, Special issue: 'Ontology of Finance' (eds. Smith, B.—Sansò, G.)*, LXIII:84, 7–24*.
Ippoliti, E. (2024). Do predictions destroy predictability? A study focusing on stock markets. *Logic Journal of IGPL*, 1–19. https://doi.org/10.1093/jigpal/jzae091
Ippoliti, E. (2024b). *Crash point. Un'analisi filosofica e pratica per capire le crisi finanziarie*. Rubbettino.
Irvine, W. B. (1987). Insider trading: An ethical appraisal. *Business and Professional Ethics Journal, 6*(4), 3–33.
James, A. (2012). *Fairness in practice: A social contract for a global economy*. Oup Usa.
Kay, J. (2024, September 5). We are all capitalists now. *Financial Times*. https://www.ft.com/content/dbc3a391-6e61-4df4-939c-cdc53a4a9c85
MacKenzie, D. (2003). An equation and its worlds: Bricolage, exemplars, disunity and performativity in financial economics. *Social Studies of Science, 33*(6), 831–868.
MacKenzie, D. A. (2006). *An engine, not a camera: How financial models shape markets*. MIT Press.
MacKenzie, D. (2017). On invoking "culture" in the analysis of behavior in financial markets. In K. Chemla & E. F. Keller (Eds.), *Cultures without culturalism: The making of scientific knowledge*. Duke University Press.
MacKenzie, D., & Millo, Y. (2003). Constructing a market, performing theory: The historical sociology of a financial derivatives exchange. *American Journal of Sociology, 109*(1), 107–145. https://doi.org/10.1086/374404
MacKenzie, D., & Spears, T. (2014). "The formula that Killed Wall street": The Gaussian copula and modelling practices in investment banking. *Social Studies of Science, 44*(3), 393–417. https://doi.org/10.1177/0306312713517157
Markowitz, H. (1952). Portfolio selection. *The Journal of Finance, 7*(1), 77–91.
Marti, E., & Gond, J.-P. (2018). When do theories become self-fulfilling? Exploring the boundary conditions of performativity. *Academy of Management Review, 43*(3), 487–508. https://doi.org/10.5465/amr.2016.0071

Mejia, S. (2022). Fiduciary duties and moral obligations to third parties; a reflection through shareholder primacy. In T. Talaulicar (Ed.), *Research handbook on corporate governance and ethics*. Edward Elgar.
Morgan, M. S. (2012). *The world in the model: How economists work and think.* Cambridge University Press.
Pardo-Guerra, J. P. (2019). *Automating finance: Infrastructures, engineers, and the making of electronic markets*. Cambridge University Press.
Sandberg, J. (2008). *The ethics of investing. Making money or making a difference?* Acta Universitatis Gothoburgensis.
Sandberg, J., & Warenski, L. (2023). *The philosophy of money and finance*. Oxford University Press.
Sansò, G. & Smith, B. (2023). Ontology of Finance. *Rivista di Estetica, 84* (3/2023), LXIII.
Svetlova, E. (2012). On the performative power of financial models. *Economy and Society, 41*(3), 418–434.
Van der Heide, A. (2019). Model Migration and Rough Edges: British Actuaries and the Ontologies of Modelling. *Social Studies of Science, 0306312719893465.* https://doi.org/10.1177/0306312719893465
Varzi, A. C. (2008). Che cos'è un derivato? Appunti per una ricerca tutta da fare. In A. Berrini (Ed.), *Le crisi finanziarie e il "Derivatus paradoxus"* (pp. 143–171). Editrice Monti.
Vergara-Fernández, M., Heilmann, C., & Szymanowska, M. (2023a). Describing model relations: The case of the capital asset pricing model (CAPM) family in financial economics. *Studies in History and Philosophy of Science, 97*, 91–100. https://doi.org/10.1016/j.shpsa.2022.12.002
Vergara-Fernández, M., Heilmann, C., & Szymanowska, M. (2023b). Contextualist model evaluation: Models in financial economics and index funds. *European Journal for Philosophy of Science, 13*(1), 6. https://doi.org/10.1007/s13194-022-00506-5
Walter, C. (2016). The financial logos: The framing of financial decision making by mathematical modelling. *Research in International Business and Finance, 37*, 597–604.

Chapter 2
How AI Can Reshape Market Efficiency

Paolo Barucca and Flaviano Morone

Abstract The efficient market hypothesis (EMH) famously stated that prices fully reflect the information available to traders. This critically depends on the transfer of information into prices through trading strategies. Traders optimise their strategy with models of increasing complexity that identify the relationship between information and profitable trades more and more accurately. The increased availability of low-cost universal approximators, such as AI systems, should be naturally pushing towards more advanced trading strategies, potentially making it harder and harder for inefficient traders to profit. In this paper, we leverage on a generalised notion of market efficiency, based on the definition of an equilibrium price process, that allows us to distinguish different levels of model complexity through investors' beliefs, and trading strategies optimisation, and discuss the relationship between AI-powered trading and the time-evolution of market efficiency. Finally, we outline the need for and the challenge of describing out-of-equilibrium market dynamics in an adaptive multi-agent environment.

2.1 Introduction

According to many economists, markets are the most efficient mechanism for collecting, aggregating, and condensing widely diverse information into a single estimate expressed by the price (Black 1986). Markets do that with trading. In financial markets, traders collectively shape price dynamics by mapping their information into trades. If many traders agree that an asset is worth more than what it is selling for, they will try to buy it, thus collectively pushing the price up and

P. Barucca (✉)
University College of London, London, UK
e-mail: p.barucca@ucl.ac.uk

F. Morone
New York University, New York, NY, USA
e-mail: fm2452@nyu.edu

E. Ippoliti et al. (eds.), *Philosophy and Finance: Ten Open Questions*, Synthese Library 504, https://doi.org/10.1007/978-3-031-96510-4_2

eliminating the potential arbitrage. In doing so, traders' individual judgements, and the information those judgements are based on, become 'priced in the market'. Thus, the process of seeking a profit ends up enhancing the efficiency of financial markets.

The theory of market efficiency (Fama 1970), in its original strong formulation, hypothesizes that prices reflect all the available information at all times and that this should imply that price changes will be independent and identically distributed, and under specific conditions follow a random walk (Bachelier 1900; Samuelson 2016). This original hypothesis did not explain what happens when new information is incorporated, as it implied an instantaneous transmission of new information into price changes. This information transmission is paradoxical: price changes only occur as a consequence of a trade, a trade that should be based on new information not reflected in the price until it occurs. Hence, informed trades should exist in order for the price to adjust to new information (Grossman & Stiglitz 1980).

Despite its paradoxical nature, the theory of efficient markets does capture a fundamental understanding of market functioning in finance, as it gives a framework to discuss the connection between the (im)possibility of traders to make a profit based on the available information and the properties of price time series. For example, as shown in Delbaen and Schachermayer (1994), the absence of arbitrage opportunities should lead to the existence of an equivalent local martingale measure for the price process.

The different original formulations of the theory of market efficiency (Fama 1970), distinguish between discrete levels of information, i.e. public and private, but do not clarify what are the implications for a trader operating in mixed marketplaces where information is unequally distributed and traders use models with different levels of intelligence, i.e. with a different ability to extract signal from a given information set. Furthermore, the hypothesis is largely independent on the ability of traders to make new trades, due to endogenous constraints, such as dry powder, or on the market impact of large investors, but also exogenous factors, such as high yield rates, and thus putting the *ability* to trade on an equal footing with the *willingness* to trade.

In this paper, we consider recent advances in the understanding of market efficiency, such as Timmermann and Granger (2004) and Jarrow and Larsson (2012), which account for trading strategies, information asymmetry and model complexity, and use it to formulate a series of hypotheses on the expected evolution of (in)efficient markets in the presence of a growing fraction of AI traders. Previous works have discussed, also critically, the impact of the whole algorithmic trading and automation on market efficiency (Yadav 2015), and only recently scholars have directly addressed the question of the specific role of AI in shaping market efficiency (Marwala et al. 2017; Mancuso 2022). Here, leveraging on a consolidated mathematical framework used to model market efficiency, we are able to discuss different levels of market efficiency and discuss why low-cost universal approximator, such as AI systems, constitute a potential novelty in the evolution of market efficiency. We hypothesize that the deployment of AI models which are able to train larger and larger classes of models at lower optimisation costs should force traders to follow higher standards of efficiency while minimizing the use of human-defined trading strategies.

Finally, we discuss the role of population dynamics in the emergence of market efficiency and the necessity of detailed agent-based models and game-theoretical analysis to understand the functioning of financial markets, especially in the presence of AI-human interactions.

2.2 Literature Review

Market efficiency is a fundamental concept in both finance and economics and has been investigated from many perspectives and with different levels of mathematical rigor. In this study, we mainly focus on the direct ramifications of the seminal work on the efficient market hypothesis by Fama (1970), which focuses on the relationship between the information available to traders, their ability to make predictable profits, and how this is reflected in the statistical properties of financial returns.

Famously, the efficient market hypothesis has three formulations: the weak, semi-strong, and strong one. In the weak formulation, the statement is the following: no trader is able to make a predictable profit solely based on available public market information. Its semi-strong formulation reads: no trader is able to make a predictable profit based on all the available public information (market and beyond). Finally, its strong formulation states: no trader is able to make a predictable profit based on both public and private information, i.e. no predictably profitable trading strategy is possible in a strongly efficient market. In other words (Fama 1970), "The strong form tests of the efficient markets model are concerned with whether all available information is fully reflected in prices in the sense that no individual has higher expected trading profits than others because he has monopolistic access to some information." This notion was initially, and still is at times, considered as an indication that neither technical nor fundamental analysis or even insider trading, can allow traders to achieve returns greater than those that could be obtained by a random buy and hold portfolio with the same level of risk (Malkiel 1989).

Market efficiency, as a general concept that hypothesizes collective forecasting consequences on price time series due to the strategies of traders, was expanded and adapted by Timmermann and Granger (2004). In this work, the authors introduced a model-specific definition of market efficiency, where price changes are independent and unpredictable conditioned on the forecasting model that the traders are using to map the available information into predictions: in an efficient market there exists an entire set of models for which the expectation of future discounted return is zero. Moreover, the authors introduce a notion of time locality in market efficiency, i.e. the possibility for markets to be temporarily inefficient and then become efficient, and the other way around. In fact, the evolutionary dynamics of market participants and trading strategies reflects the adaptive nature of market efficiency (Lo 2005), and poses the challenge of modelling out-of-equilibrium market dynamics, i.e. how financial markets approach efficiency, or fall out from it.

In Lim and Brooks (2011), the authors observe that the possibility of weak-form time-varying market efficiency has received increasing attention in the last decades, showing as evidence the time-varying structure of autocorrelations measured in Ito and Sugiyama (2009). Although accounting for model specification, the model of market efficiency by Timmermann and Granger (2004) is effectively limited by the need to introduce an arbitrary pricing kernel (*viz.* pricing model) to actually conduct a forecasting test over the returns. The need to assume a specific asset pricing model, the so-called "joint hypothesis problem", was tackled by the seminal paper (Jarrow & Larsson 2012), where the authors combine the definition of an economy, consisting of a set of investors trading on a market, and define market efficiency as the ability of the financial market price to converge towards an equilibrium price process, thus avoiding the need to impose an asset pricing model.

Real markets cannot be regarded as such efficient aggregators of information that no trader should ever be so naive as to hope to beat them. Traders make mistakes, markets are not perfect, and cannot be regarded as strongly efficient, as noted early in the literature (Gordon &Kornhauser 1985), and stated Fama himself, "We would not, of course, expect this model to be an exact description of reality, and indeed, the preceding discussions have already indicated the existence of contradictory evidence." However, experience reveals that even if markets are less efficient than the strong form of the hypothesis supposes, it is still very difficult to consistently beat them, which is why so few traders can allegedly do it. Therefore, we can consider these definitions as ideal scenarios serving as a framework for developing tests for quantifying real market efficiency.

The notion of market efficiency as prices fully reflecting information can be interpreted in two main ways: (i) statistical unpredictability of the price time series and (ii) prices ought to be adherent to a fundamental value that the available information should entail (Black 1986).

These two interpretations, the statistical and the fundamental one, differ substantially in the ways they can be tested: whilst predictability of price time series can be formulated in terms of expectation over future profits, potentially including discount factors, dividends and pricing kernels (Walter 2006; Timmermann & Granger 2004), adherence to a fundamental value supposes the ability to build a structural model for the price formation based on the information available to traders.

Measuring efficiency based on the discrepancy between asset prices and their fundamental values is difficult for one main reason: even if there were a structural model for fundamental the value of an asset, it would need to take into account the actual market where the asset is traded and the prices that traders would be willing to pay for it, hence it would need to model the market itself. Unsurprisingly, tests for market efficiency have focused on the statistical interpretation of the efficient market hypothesis, mainly testing the ability of factor models to predict price returns (Fama et al. 1969; Fama & French 1993). This major limitation was addressed in Jarrow and Larsson (2012), where the authors introduce an economy associated with the market and define market efficiency as the correspondence between the security market price process and the equilibrium price process (Duffie 1986), of a commodity economy. The authors avoid the "joint hypothesis problem": the necessity of testing market efficiency within a given equilibrium model. Their definition reconciles the statistical and fundamental interpretation of market efficiency by showing the correspondence between the notion of no-arbitrage (Delbaen & Schachermayer 1994), in the security market, and the existence of an equilibrium price process for the underlying commodity market. We will mainly use the formalism and equilibrium notions introduced in Jarrow and Larsson (2012) to discuss the out-of-equilibrium evolution of market (in)efficiency.

2.3 Market Efficiency, Investors' Beliefs, and Trading Strategies

The efficient market hypothesis models efficiency in terms of the statistical unpredictability of the return time series that derives from the necessity of the absence of arbitrage opportunities. It starts from an idealised view of financial markets where traders have access to the same information set (Fama 1970), and from that it hypothesizes that markets with informed traders should display randomly fluctuating returns with no arbitrage opportunities, or, more generally, efficiency should imply the existence of an equivalent martingale measure for the price process.

The information set is a cumulative collection of time-structured data, including traditional macroeconomic time series, financial indicators, alternative data, high-resolution market data, fundamental data about individual stocks, firms, sectors, and more. In Jarrow and Larsson (2012) the information available to traders is expressed as a filtration over a probability space. The investors' beliefs in Jarrow and Larsson (2012) represent the level of model capacity the investors have achieved, i.e. they constitute a form of market intelligence. In this definition of market efficiency, investors' beliefs are crucial as they need to be equivalent to the measure of the probability space. In Jarrow and Larsson (2012) the probability measure defines the price process, which is defined as efficient if it corresponds to the equilibrium price process of a security and commodity market. In this setting, model complexity is embedded in the probability measure and market efficiency implies the existence of an equivalent martingale measure. Both these generalised definitions allow us to consider market (in)efficiency not only at a global market level but at a trader's level. In particular, in Jarrow and Larsson (2012), the economy associated with the market consists of a finite set of investors each provided with their beliefs, information, and utility functions. Beliefs, information, and utility functions all contribute to shape the individual investor's problem and consequently their optimal consumption and trading strategy. In this framework, in the presence of market efficiency, all the individual optimal trading strategies are maximal, as demonstrated in Jarrow and Larsson (2012), and, notably, are dominated by a simple market portfolio (and by each security holding). Given any market, to disprove efficiency one simply needs to find an arbitrage opportunity or a trading strategy dominating the market portfolio. Arbitrage opportunities can emerge when the assumptions that lead to a market equilibrium are not met and investors fail to identify an optimal solution to the investor's problem, so that their consumption choices and the trading strategies are not optimal, and the market prices do not necessarily provide an equilibrium price process for the economy.

2.3.1 From Efficient Markets to Efficient Traders

The efficient market in Jarrow and Larsson (2012) is associated with an equilibrium price process which embodies the structural model that should map all the available information into prices; all traders have access to the available information and their beliefs are equivalent to the price probability measure. In real markets, traders may differ on the information they can gather, their beliefs' and on the trading strategies they can build. Formally, we have two separate objects: the complete filtered probability space of the market, and the investors' beliefs with their information.

We then proceed to classify different investors in the market based on their relative ability to detect signal in the market, i.e. to align their beliefs to the probability measure of the information set. In practice, this is an inference problem for traders who need to update their beliefs based on the observed information and price processes.

Let us assume a market where the set of available information is the same for all traders and coincides with the set of information of the market itself. We can now consider two main cases:

- traders who are unable to align their beliefs with the probability measure of the process, and to find an optimal solution to the investor's problem (inefficient traders),
- traders who are able to align their beliefs with the probability measure of the process and find an optimal solution to their investor's problem (efficient traders).

In this scenario, there is no guarantee for market efficiency, and as a consequence the no-dominance does not apply, i.e. there could be profitable admissible strategies that outperform the market portfolio, and we expect the efficient traders to be the ones to identify them.

Efficient traders need both to identify the right probability measure and the optimal solution to the investor's problem. This comes at a cost, that we name optimization cost, defined as the cost needed to discover the probability measure. Further, even if the true probability measure became available to efficient traders, they still may not be able to identify the optimal strategies. If only some traders succeed, then the market is inefficient, there exists a non-empty set of traders who can make predictably profitable trades, i.e. the market can exhibit arbitrage opportunities. In inefficient markets, efficient traders can beat the collective intelligence of the market. For a simple isolated small slow human investor using limited information, i.e. an uninformed simple trader, in an inefficient market, it should be almost impossible to do better than efficient traders. In inefficient markets competition is key and no trader can simply assume that a simple market portfolio will not be outperformed (on average) by better strategies. The emergence of market efficiency is positive especially for uninformed traders, i.e., when market efficiency applies, uninformed traders can simply invest in the market portfolio and forget about complex tasks such as aligning their beliefs and identifying complex (and expensive) trading strategies. This is a paradoxical perspective provided by market efficiency: it eliminates the out-of-equilibrium dynamic competition that is the very reason market efficiency should emerge.

2.3.2 Co-evolution of Price and Beliefs

In real (in)efficient markets, we still expect arbitrage opportunities to disappear, i.e. inefficient markets should approach efficiency: we expect the distance between the equilibrium price process and the realised price process to remain small. As discussed in Timmermann and Granger (2004), "An efficient market is thus a market in which predictability of asset returns, after adjusting for time-varying risk-premia and transaction costs, can still exist but only 'locally in time' in the sense that once predictable patterns are discovered by a wide group of investors, they will rapidly disappear through these investors' transactions." In other words, predictable patterns, when they exist, they tend to self- destruct after a certain period of time. Arguably, this is how the collective intelligence of markets should emerge: the more traders expand their information sets and improve their capacity of estimating beliefs, the more the set of probability measures should evolve and push the equilibrium price process to higher complexity levels, where by complexity we here refer to a measure of the intricacies of the layered non-linear multivariate dependencies between the variables of the system.

The presence of an arbitrage opportunity could be associated to information asymmetry and to the ability of some investors to better identify optimal trading strategies thanks to their better information gathering and belief estimation. This fundamental mechanism creates a pressure for the time evolution of the equilibrium price process in efficient markets: the market price cannot be indifferent to investors' beliefs even when they are potentially far from the market probability measure as this would lead to the persistence of arbitrage opportunities. This could be the case if investors were systematically unable to gather relevant information for the price process or to approximate the probability measure, due to modeling limitations. In this setting, for the market to be efficient it would need to temporarily co-evolve with the traders' beliefs, i.e. the probability measure would need to be aligned with the investors' beliefs and not the other way around, at least until traders were able to update their information and improve their modeling capacity. In particular, here we argue that the probability measure complexity should co-evolve with the complexity of the investors' beliefs in the market. In an evolving close-to-efficiency inefficient market, the probability measure should evolve over time, so that its complexity remains close but above the average complexity of the beliefs in the market.

So far we have considered traders with different beliefs' but sharing the same information. In general, as originally discussed in Fama (1970), we can distinguish different information sets, for example market information on price time series, public information on companies, and private information on business strategies and more. The acquisition and integration of information in the models available to traders and in the equilibrium price process are a fundamental part of the market evolution. In Jarrow and Larsson (2012), the cases of information reduction and information expansion are considered: information reduction, namely the reduction of the filtration in the filtered probability space, is found to preserve market efficiency whilst information expansion is generally not.

The probability measure may be based on a large information set and traders may have access to limited information sets that hinder their ability to design profitable strategies. In contrast, the acquisition of new information by traders could lead to the identification of new equilibrium price processes depending on larger information sets. The set of data that are now available about companies and their time resolution has significantly increased and, at the same time, this requires an increase in the models' ability to incorporate this information and translate that into optimal predictions.

The cost of collecting and processing this increasing amount of data pushes traders to improve their ability of extracting signal from them, i.e. a large investment on data collection needs to be coupled with an investment in the model complexity, at least up to a certain point, to make sure that all the trading signal that is present in the data is identified and exploited. This implies that the investors' beliefs and the equilibrium price process should co-evolve with the set of available information. More information requires more complex models, able to extract new signals from the new information; conversely the acquisition of new information pushes traders to improve their beliefs', which as a consequence should push the equilibrium price process to higher complexity levels.

2.4 Low-Cost Universal Approximators as Drivers of Market Efficiency

The time evolution of market efficiency has been investigated mainly through the study of the existence of regime changes and structural changes in autoregressive models (Lim & Brooks 2011). Linear models have the advantage of efficient and reliable solvers to identify the optimal models, but only consider linear combinations of factors to determine the price. Universal approximators have been known and studied in functional analysis for a long time (Stone 1948), and recently multilayer perceptrons (Hornik et al. 1989; Barron 1993), and their generalisations, have been recognised and widely utilised as such. In particular, in Hornik et al. (1989), the authors demonstrated that multilayer feedforward networks are capable of approximating any Borel-measurable function from one finite dimensional space to another to any desired degree of accuracy, provided sufficiently many hidden units are available, and multiple other studies have extended and generalised this result (Cybenko 1989; Barron 1993).

The ability of approximating a function with a class of other basis functions is not a novel result, yet the limited generalisation errors and the efficiency of the training algorithms, yielding relatively low computational costs, for neural networks (Rumelhart et al. 1986), have notoriously established the use of these models in a wide range of applications. In finance, a growing number of studies is showing how AI algorithms can outperform traditional statistical models, e.g. Fischer and Krauss (2018) and Makridakis et al. (2020), and multiple independent surveys on financial firms have recognised that a vast majority of traders and investors (Jung et al. 2019), will be relying more and more on AI models for their financial decisions.

The ability of AI models of acting as universal function approximators, also in presence of different modalities of data, such as text, time-series, images, makes them suitable models to adopt in the space of available models for traders. Despite the existence of other universal approximators, AI models are unique in terms of multimodality, capacity, and adaptability. Linear approximators are generally not effective with multimodal data, not easy to extend in terms of model capacity (unless they take the form of AI models themselves), and less flexible in terms of available architectures and optimisation schemes. Universal approximation theorems guarantee that, provided the convergence of the training algorithm, with sufficient data all informed AI traders will never under-perform other traders and, if existent, should always be able to identify and approximate the probability measure of the equilibrium price process. We argue that, under these hypotheses, a market of rational traders will be inclined to adopt AI models and, as discussed above, the equilibrium price process should be pressured to become AI-efficient, i.e. not even AI models should, eventually, be able to identify profitable strategies. We may refer to this as asymptotic AI-efficiency. At that stage, the competition in the market will focus entirely on the rapid integration of new information, rather than on the model itself. Convergence and big data are not guaranteed, so models will keep changing in terms of data available and players in the market.

2.4.1 Trading in a World of AI-Efficient Traders

The notion of economy-dependent market efficiency (Jarrow & Larsson 2012), is extremely useful to provide a richer picture of economic equilibrium in financial markets, yet it leaves us with complex questions over markets' evolution: how does the space of available information evolve? And, how do beliefs and price process co-evolve?

Hardware and software design has led to an increased capability to harness information and to an increased ability to explore the space of statistical models for estimating beliefs and developing trading strategies. The specific way in which the price process varies over time depends on the market fundamentals, market design, market microstructure, e.g. the liquidity of the order book, and ultimately on the traders themselves.

In a realistic setting, traders share a certain fraction of information, as well as certain classes of models, nevertheless it is also likely that each trader will have a slightly -or significantly- different set of information and, at the same time, a different class of models that they are able to use. Hence, we should look at the changes in market (in)efficiency in two main directions: one is a longitudinal time dimension, i.e. information and beliefs change and evolve over time; and another one is cross-sectional, i.e. in a given market at a given time multiple traders can coexist who do not share the same information and the same beliefs.

The absence of profitable trades is hypothesized as a consequence of the market ability to identify perfect prices based on the information available to different traders, and this efficiency is enabled by the complexity of the equilibrium price process that is accounting for the collective intelligence of the market.

So far we have made the simplifying assumption that the equilibrium price process at a given time is a function of the information set which is independent from the beliefs and strategies of individual traders. In reality markets do display some form of co-dependence and reflexivity: the performance of a strategy will not only be based on its expected value but also on the simultaneous strategies of the other traders, e.g. the specific sequence in order placement can significantly affect the price that a trader can obtain with respect to others. The dynamic time-dependent mapping between information and optimal trades will depend on the simultaneous, or quasi-simultaneous, decisions of the other traders, that will partially move the optimal solution with each of their trades. This can be partially understood in a game-theoretical way. A well-known example is the following. Consider a game where players (traders) must guess a number between 0 and 100 and the winner is the player who guesses the number which is closest to $2/3$ of the average guess. The Nash equilibrium, i.e. the solution to this game, is 0, so in theory it is impossible to beat the crowd (the market) if every player uses the optimal strategy. But real markets are not perfect as we said. In fact, this contest was actually run by the *Financial Times* in 1997 and the result was that, although many contestants were able to figure out the Nash equilibrium and guessed 0, they were wrong in thinking that everyone else would be as smart as they were (Thaler 2016). In fact the average guess was 18.9 and so the winning guess was 13. This experiment thus shows that the performance of a strategy in a real-market depends on the simultaneous strategies of the other traders and so to beat the market a trader must model also all other traders' behavior.

In the framework we presented for modeling efficiency, we essentially have two rate of changes in complexity, the change in market complexity, that is how rapidly the complexity of the probability measure changes over time and how rapidly the complexity of the beliefs' models available to traders change.

Traders' directly affect market efficiency and they compete with the market price dynamics they actually contribute to define. Market efficiency is related to the models available to the traders who populate it. Let us, for instance, consider a market populated with traders with limited information and limited model capacity. It is likely that a new trader with more information and more model capacity could enter the market and easily find profitable strategies, i.e. the market would be inefficient. The precise mechanism by which the traders' ability is translated into the collective intelligence of the market is generally dependent on the details of the market design.

This complex adaptation makes financial markets different from a static and predictable computational task, such as classifying pictures or even solving combinatorial optimisation problems, and constitutes the very reason a market of investments or hedge funds exists: profitable trades are the consequence of a powerful prediction model and a successful behavioral agent-based model, Bouchaud (2018). Traders are both responsible for the emergence of the collective intelligence of the markets whilst being in competition with it and, as agents, are effectively following a noisy learning algorithm trying to optimise their different objective functions in a non-stationary multi-player environment.

2.5 Conclusions

In this paper, we discussed the notion of complexity in market efficiency and the importance of its out-of-equilibrum dynamics. Introducing different levels of complexity can allow market observers to describe the evolution of market (in)efficiency over time and make sense of the ability of traders to adapt to changing market conditions. More complex models are now available to traders to map the information into beliefs and beliefs into trading strategies, potentially leading to a form of market singularity, i.e. where all traders can effectively learn accurate prices instantaneously. We discussed how the existence of universal approximators and their increasing availability and decreasing computational costs could affect the evolution of market efficiency. On the one hand, low-cost AI approximators could push the boundary of market efficiency to a highly competitive global AI-efficiency. However, models are limited by convergence, computational power, and by the complex modeling of interactions between strategies. At the same time, we are observing the complexification of the economic dependencies, of the interconnectedness of countries, firms, and supply networks, the increase in the responsiveness of consumers' preferences, which can all contribute to hinder the predictability of the market performance of different stocks and companies, so that even more information and more complex models will be needed to reach a given market performance, and markets may complexify at a rate not even AI-traders will be able to adapt to.

References

Bachelier, L. (1900). Théorie de la spéculation. In *Annales scientifiques de l'École normale supérieure* (vol. 17, pp. 21–86).

Barron, A. R. (1993). Universal approximation bounds for superpositions of a sigmoidal function. *IEEE Transactions on Information Theory, 39*(3), 930–945.

Black, F. (1986). Noise, 41j. *Fin, 529*, 533.

Bouchaud, J. P. (2018). Agent-based models for market impact and volatility. *Handbook of Computational Economics, 4*, 393–436.

Cybenko, G. (1989). Approximation by superpositions of a sigmoidal function. *Mathematics of Control, Signals and Systems, 2*(4), 303–314.

Delbaen, F., & Schachermayer, W. (1994). A general version of the fundamental theorem of asset pricing. *Mathematische Annalen, 300*(1), 463–520.

Duffie, D. (1986). Stochastic equilibria: existence, spanning number, and theno expected financial gain from trade'hypothesis. *Econometrica: Journal of the Econometric Society, 54*, 1161–1183.

Fama, E. F. (1970). Efficient capital markets: a review of theory and empirical work. *The Journal of Finance, 25*(2):383–417.

Fama, E. F., & French, K. R. (1993). Common risk factors in the returns on stocks and bonds. *Journal of Financial Economics, 33*(1), 3–56.

Fama, E. F., Fisher, L., Jensen, M. C., & Roll, R. (1969). The adjustment of stock prices to new information. *International Economic Review, 10*(1), 1–21.

Fischer, T., & Krauss, C. (2018). Deep learning with long short-term memory networks for financial market predictions. *European Journal of Operational Research, 270*(2), 654–669.

Gordon, J.N., & Kornhauser, L.A. (1985). Efficient markets, costly information, and securities research. *NYU Law Review, 60*, 761.

Grossman, S. J., & Stiglitz, J. E. (1980). On the impossibility of informationally efficient markets. *The American Economic Review, 70*(3), 393–408.

Hornik, K., Stinchcombe, M., & White, H. (1989). Multilayer feedforward networks are universal approximators. *Neural Networks, 2*(5), 359–366.

Ito, M., & Sugiyama, S. (2009). Measuring the degree of time varying market inefficiency. *Economics Letters, 103*(1), 62–64.

Jarrow, R. A., & Larsson, M. (2012). The meaning of market efficiency. *Mathematical Finance: An International Journal of Mathematics, Statistics and Financial Economics, 22*(1), 1–30.

Jung, C., Mueller, H., Pedemonte, S., Plances, S., & Thew, O. (2019). Machine learning in UK financial services. *Bank of England and Financial Conduct Authority*.

Lim, K.-P., & Brooks, R. (2011). The evolution of stock market efficiency over time: A survey of the empirical literature. *Journal of Economic Surveys, 25*(1), 69–108.

Lo, A. W. (2005). Reconciling efficient markets with behavioral finance: the adaptive markets hypothesis. *Journal of Investment Consulting, 7*(2), 21–44.

Makridakis, S., Spiliotis, E., & Assimakopoulos, V. (2020). The m4 competition: 100,000 time series and 61 forecasting methods. *International Journal of Forecasting, 36*(1), 54–74.

Malkiel, B. G. (1989). Efficient market hypothesis. In *Finance* (pp. 127–134). Springer.

Mancuso, F. (2022). The efficient market hypothesis and trading ai advancements: an overview. https://tesi.luiss.it/33475/1/243261_MANCUSO_FRANCESCO.pdf

Marwala, T., Hurwitz, E., Marwala, T., & Hurwitz, E. (2017). Efficient market hypothesis. *Artificial Intelligence and Economic Theory: Skynet in the Market* (pp. 101–110). Springer.

Rumelhart, D. E., Hinton, G. E., & Williams, R. J. (1986). Learning representations by back-propagating errors. *Nature, 323*(6088), 533–536.

Samuelson, P. A. (2016). Proof that properly anticipated prices fluctuate randomly. In *The world scientific handbook of futures markets* (pp. 25–38). World Scientific.

Stone, M. H. (1948). The generalized weierstrass approximation theorem. *Mathematics Magazine, 21*(5), 237–254.

Thaler, R. H. (2016). Behavioral economics: past, present, and future. *American Economic Review, 106*(7), 1577–1600.

Timmermann, A., & Granger, C. W. J. (2004). Efficient market hypothesis and forecasting. *International Journal of Forecasting, 20*(1), 15–27.

Walter, C. (2006). Les martingales sur les marchés financiers. *Revue de synthèse, 127*(2), 379.

Yadav, Y. (2015). How algorithmic trading undermines efficiency in capital markets. *Vanderbilt Law Review, 68*, 1607.

Chapter 3
A Tour Into Dark Pools

Sühan Altay, Katia Colaneri, and Zehra Eksi-Altay

Abstract Dark pools are alternative trading venues that obscure much information about trades. Initially used by institutional investors to execute large orders discreetly, various types of dark pools exist today for different purposes beyond reducing price impact. This paper provides a historical overview of dark pools, inquiring about the reasons for their existence as well as their impact on financial markets. We also discuss the ethical and moral implications, including fairness and transparency.

Keywords Dark pools · Informed trades · Price discovery · HFT · Regulations · Ethical issues

3.1 Introduction

Dark pools are private trading platforms distinct from public stock exchanges (or *lit-pools*) and they are designed to allow investors to trade substantial stock volumes without revealing their intentions, thereby avoiding significant market price shifts. These venues prioritize confidentiality, ensuring large order details remain undisclosed, especially before execution. Notably, they execute trades within the bid and ask spread of public exchanges, often securing better prices for both parties and typically having lower associated fees than their public counterparts.

S. Altay
Institute of Statistics and Mathematical Methods in Economics, TU Wien, Vienna, Austria

K. Colaneri (✉)
Department of Economics and Finance, University of Rome Tor Vergata, Roma, Italy
e-mail: katia.colaneri@uniroma2.it

Z. Eksi-Altay
Institute for Statistics and Mathematics, WU-Vienna University of Economics and Business, Vienna, Austria
e-mail: zehra.eksi@wu.ac.at

E. Ippoliti et al. (eds.), *Philosophy and Finance: Ten Open Questions*, Synthese Library 504, https://doi.org/10.1007/978-3-031-96510-4_3

Conventional exchange venues, or standard *lit* pools, such as the New York Stock Exchange (NYSE), the Nasdaq, the Euronext, or the London Stock Exchange, are characterized by displayed (buying and selling) orders. All participants of a lit market have access to the order book information, that is, they know orders volume,[1] prices, and a stock's liquidity level.[2] These valuable pieces of information can be used to infer stock price movements in the short term and eventually redirect investment decisions. For this reason, specific categories of investors, e.g., institutional investors, whose trading choices may influence the market, find it particularly helpful to hide their intentions to avoid, for instance, *predatory trading*, as illustrated, e.g., in Brunnermeier and Pedersen (2005).

To better illustrate the problem, consider the following example. Suppose that an investor, for instance, an investment bank, is attempting to sell a large number of shares on a public exchange. When a significant selling block appears on the order book in a visible trading venue, institutional investors and high-frequency traders swiftly respond, for example, placing a buy order with a lower ask price, causing a drop in the security's value before the bank can secure enough buyers to complete the entire order at the initial price. This phenomenon is called the *market price impact* (see, e.g., Bouchaud 2010). In contrast, if the trade details are disclosed only after the completion of an order, the market would be less affected, as the trade has already occurred. This example is aimed to underline the power of information and to smoothly drive the reader into a different market environment, whose rules substantially permit keeping investors' intentions hidden.

Dark pools cater primarily to institutional investors, including pension funds, mutual funds, hedge funds, banks, and private equity funds. What distinguishes dark pools from standard trading venues is their provision of anonymous trading, designed to minimize or even eliminate the adverse price effects (e.g., market price impact) commonly associated with the large-scale trades in the standard lit markets.[3] Typically, participants in dark pools receive more favorable quotes, often hovering around the midpoint between the best bid and best ask prices. Notably, the pre-trade information regarding orders executed within these venues remains concealed from the public and other market participants. Accordingly, the term *dark pool* refers to the lack of pre-trade transparency in these venues. Only after the trades have concluded, details regarding successfully executed orders become publicly accessible.

The first type of dark pools dates back to the end of the 1970s, and since then, they have experienced a progressive increase in the number of transactions executed. Over the years, the nature of the dark pools has changed more recently by

---

[1] Volume refers to the number of shares of a security traded in a given period of time.

[2] Liquidity stands for how easy is to sell or buy a certain security without substantially affecting the asset's price; the easier the transaction takes place, the more liquid the security. See, e.g. Hintz (2014, Page 332).

[3] For how to trade optimally in lit markets, please see the pioneering works by Bertsimas and Lo (1998) and Almgren and Chriss (2001).

the initiation of the so-called *second generation dark pools*, or *lit-dark pools* (for example, the internalized pools) mainly managed by big investment banks. Further classifications of the dark pools are discussed, for instance, in Mittal (2008, Section: Taxonomy of Dark Pools).

While trading in a dark pool offers notable advantages in terms of price execution, particularly for large investors, concerns have emerged over time. These concerns revolve around market transparency, fairness, and potential conflicts of interest and have become more and more controversial with the emergence of high-frequency trading (HFT) in these venues. This discussion was also brought to the public by Lewis (2014), who argues that the opacity of dark pools creates an uneven playing field, favoring high-frequency traders who can exploit the lack of transparency to gain an informational advantage. HFT firms use sophisticated algorithms and high-speed connections to detect and front-run large orders placed in dark pools, profiting from the price movements that result from these trades, hence challenging basic rules like market fairness and transparency.

As a result, regulatory bodies such as the Securities and Exchange Commission (SEC) and Financial Industry Regulatory Authority (FINRA) in the United States, along with the European Securities and Markets Authority (ESMA) in Europe, have implemented regulations to tackle these issues.

Academic contributions aimed at describing the characteristics of dark pools and the underlying advantages and disadvantages of trading in these venues include but are not limited to Brugler (2015), Zhu (2014), Shorter and Miller (2014), Hatch (2009), Cooper et al. (2016), Cartea et al. (2015), Brolley (2020), and Buti et al. (2022).[4]

This paper discusses dark pools' basics and historical evolution, investigating specific aspects of hidden information dissemination and its effects on market behavior. It also explores certain ethical dimensions of dark pools, including questions about fairness, transparency, and market efficiency. Moreover, we examine the moral considerations and regulators' role in dark pools.

The paper is structured as follows: In Sect. 3.2, we give fundamental information about dark pools, underlining in Sect. 3.2.1 its differences with the lit market. Section 3.3 contains a detailed historical account. In Sect. 3.4 we discuss dark pools from various points of view. In particular, we examine the impact of dark pool trading on conventional market dynamics in Sect. 3.4.1, we recall the legal actions taken by regulatory authorities in Sect. 3.4.2, and we conclude the paper with moral and ethical discussions in Sect. 3.4.3.

[4] The appealing characteristics of dark pools have also attracted the attention of many researchers in quantitative finance. In this paper, we are not focusing on designing and analyzing mathematical models for trading in the dark pools. However, we should mention that the literature on this topic is quite rich and includes, for instance, Kratz (2014), Horst and Naujokat (2014), Kratz and Schöneborn (2015), Crisafi and Macrina (2016), Buti et al. (2017), Gao et al. (2018), Colaneri and De Angelis (2022), and Baldacci et al. (2023).

3.2 Basic Underlying Mechanism of the Dark Pools

Institutional investors, such as mutual, pension, and hedge funds, must register to access a dark pool. These participants are usually required to meet specific criteria and comply with regulations. Institutional investors submit their orders via the dark pool's electronic trading platform when participating. These orders often involve large quantities of shares or other financial instruments. Subsequently, the dark pool's trading platform employs algorithms to match buying orders and selling orders from various participants within the pool. The primary objective is to identify the best match to complete the trade at a price that is typically close to the prevailing market price.

Generally speaking, dark pools implement the following mechanism: the trade's execution price is determined by the market price at the moment of the trade rather than by the order's initial limit price. After a match is found, the trade is executed within the dark pool, and the trade details are not immediately disclosed to the public exchanges. Although dark pools provide a certain level of anonymity during trading, they are still subject to regulatory reporting requirements. After the execution of a trade within a dark pool, relevant trade information is typically conveyed to regulators for the purpose of transparency and monitoring.

The settlement process for trades executed in a dark pool is the same as those executed on public exchanges. Securities are transferred between the parties involved, and the funds are exchanged accordingly.

3.2.1 Lit Markets vs Dark Pools

In this section, we compare the two market venues, shedding light on their unique characteristics and functionalities. This comparison will help us discuss the impact of dark pools in financial markets later on.

In traditional stock exchanges, institutional investors faced with, for example, selling a substantial block of shares often resort to executing the sale incrementally across multiple trading sessions, possibly spanning several days. This gradual liquidation strategy frequently exerts downward pressure on the stock's market price as the market becomes aware of the presence of a substantial seller.

Conversely, in the realm of dark pools, where the order book remains private, participants enjoy a favorable environment to execute larger trades without influencing market prices. This feature offers the potential benefit of enhancing the financial outcome for the larger investor, whether through increased sales proceeds or reduced purchase costs.

Dark pools offer institutional investors a remarkable edge in terms of confidentiality and privacy. This characteristic is invaluable when handling sensitive securities or preventing market manipulation. By executing trades anonymously within dark pools, institutional investors shield their positions from the other market participants, effectively mitigating the risk of front-running and other detrimental market activities. In contrast, public exchanges operate openly, with trades visible to all participants, providing transparency but at the expense of privacy.

Dark pools can also help improve execution quality for institutional investors. This is because dark pools often provide access to liquidity that is not available on lit markets, reduced transaction costs, more favorable execution prices, and reduced commission fees.

Dark pools have limited price transparency. Unlike on public stock exchanges, where the trade price is often set by the initial limit price specified by the trader, in dark pools, the trade's price is established based on the prevailing market price at the moment the trade is executed. For example, suppose that an investor wants to buy a large block of shares at a specific limit price, say $100 each, and submit this order to a dark pool. In this case, the actual buying price may be slightly higher or lower than $100, depending on the market conditions at that precise moment in which the trade is executed. This approach helps prevent information leakage, as it does not broadcast the investor's intention to buy at a specific price to the broader market. This mitigates the risk of other traders adjusting their strategies in response to the order. Additionally, it curtails the impact of the trade on the overall market by avoiding sudden price swings that could occur if an investor's large order were to be executed at a significantly different price than the current market price. However, this makes it difficult to determine the fair value of securities traded in dark pools.

A trading practice that is used to circumvent the lack of information is *pinging the book*. That strategy consists of submitting small, frequent orders not for the purpose of portfolio liquidation but rather to learn about the liquidity state of a dark pool and prices, detect the names of large orders, and predict future price moves. Lately, this mechanism has been at the center of controversies, primarily related to a lack of clarity by some dark pools of their operating mechanisms. On the other hand, many pools have adopted measures such as minimum size restrictions or banned immediate-or-cancel orders[5] to offer extra protection to their clients and, hence to stem the tide of pinging.

Large traders, such as institutional traders, often gravitate towards dark pools to execute large orders discreetly, potentially limiting the information available to other market participants. In contrast, lit stock exchanges generally facilitate more transparent order books. Nevertheless, the interplay between dark pools and lit exchanges can have nuanced effects, as dark pools may concentrate on specific types of trading, potentially raising concerns about market fragmentation and transparency.

Dark pools and lit markets also diverge notably in terms of flexibility. Indeed, dark pools grant investors heightened flexibility in executing securities trades due to their exemption from the stringent reporting requirements applied to public exchanges. This flexibility extends to the ability to place limit orders within dark pools; this feature is especially advantageous when dealing with illiquid securities

---

[5] Immediate-or-cancel orders are orders that execute as much as possible at the best price and the rest are canceled, see, e.g., Cartea et al. (2015, Chapter 1, page 13).

or securities that are prone to significant price fluctuations. In contrast, lit markets adhere to strict reporting and trading regulations, which can limit the range of trading strategies and potentially expose investors to increased market impact costs.

3.3 Brief History and the Development of the Dark Pools

Over time, the popularity of dark pools grew, and by 2020, dark pools accounted for approximately 14% of all U.S. equity trading volume (see, Bayona 2020 and references therein). In Europe, the development of dark pools followed a similar trajectory, albeit at a slightly slower pace due to regulatory differences among EU member states. According to the report by ESMA (2020, page 21), in 2019, dark pool trading accounted for 9% of the total trading volume.

Below, we provide a chronological overview of critical milestones in the history of dark pools.

1979: The Securities and Exchange Commission (SEC) introduced Rule 19c-3, a regulatory change that allowed securities listed in a particular market to be traded privately, marking an early step toward developing dark pools. To understand the importance of this regulation, a bit of background is needed. In the 1980s, the main trading venues for US shares were the NYSE and its alternative, the Nasdaq. Fund-management firms could not trade directly in the NYSE but were required to trade via an NYSE member. A drawback of this situation was that high commissions were charged for trading through a member of the NYSE. Although a former NYSE's commission rule of charging at least 0.25% of the value of the transaction was abolished in 1975 after pressure from Congress and the SEC, commissions remained substantial. Moreover, fund managers were concerned about being taken advantage of by investment banks and broker-dealers. In parallel, new market devices started emerging in the 1970s, allowing fund managers to trade directly at lower costs. Instinet was a pioneering electronic trading platform allowing bids and offers with an anonymous order book. The mechanism could be summarised as follows: fund-management firms that were subscribers to Instinet could place bids for purchasing shares or offers for selling them through terminals connected to Instinet's central computer system. Any bids or offers that couldn't be executed right away were recorded in a collective, anonymous order book accessible to all subscribers on their terminal screens. In this system, the order book was accessible, hence not dark.

1986: Instinet launched the first dark pool, the Crossing Network, providing institutional investors with a discreet trading venue. The main goal of this service was to eliminate market impact. After the closure of the NYSE, fund managers had the opportunity to use the Crossing Network to anonymously submit electronic bids to buy shares or offers to sell them. These bids and offers were not disclosed to other users. Users could input the desired number of shares, and the price was set at the closing price of the shares for that day, with a commission of one cent per share of executed orders.

1987: Investment Technology Group (ITG) introduced POSIT, the first dark pool designed for intra-day order execution, expanding the range of opportunities for institutional traders seeking minimal price impact. In this venue, buy and sell orders were executed at public market prices.
1998: The SEC introduced regulatory measures known as the Regulation Alternative Trading System (ATS), marking a significant turning point for dark pools. These regulations required dark pools to register as exchanges or broker-dealers with the FINRA, subjecting them to trading rules like traditional exchanges. However, dark pools were not obligated to disclose their quotes publicly.
2001: A more sophisticated first-generation dark pool venue, Liquidnet, was launched. This system allowed for anonymous negotiations between fund managers for buying and selling shares. Liquidnet had continuous electronic access to fund managers' order management systems, facilitating efficient negotiations and transactions in an anonymous and rule-governed manner. Although very attractive first-generation pools remained marginal, mainly because the traditional exchange had higher execution rates and offered *soft-dollar* compensations.[6] A detailed analysis of this phenomenon and its impact on first-generation dark pools is provided, for instance, by MacKenzie (2019).
1990s–2000s: In the 1990s, electronic *lit* markets arose, challenging Nasdaq and NYSE. Investment banks offered algorithmic execution alongside human traders, like Credit Suisse's Advanced Execution Services. Investment banks aimed to *internalize* orders, buying from one fund manager and selling to another, saving on external market fees. On the other side, the term *internalization* gained a negative connotation due to potential fairness concerns. The adoption of the Regulation National Market System (Reg NMS) by the SEC in 2005 increased the complexity and regulation of internalization and brought critical changes in the market structure. It removed a previous rule safeguarding manually submitted exchange quotes, fostering more significant innovation in electronic trading venues. While promoting electronic trading, Reg NMS also contributed to market fragmentation. Second-generation dark pools were introduced to maintain the economic benefits of internalization for investment banks while enhancing their legitimacy. This transformation aimed at modernizing the inherent reliance of fund managers on investment banks. MacKenzie (2019) makes the following comparison between the first and second-generation dark pools:

> The rules of the earliest of the first-generation dark pools did not allow users to specify the prices at which they wanted to buy or sell. Instead, the pools automatically matched buy and sell orders at the midpoint of the price range prevailing on exchanges such as the NYSE [...]. Second-generation US dark pools more closely mimicked electronic exchanges: although orders are in practice often simply matched at the midpoint, users can, for example, specify a maximum price at which they will buy shares, and an order of this kind will be executed by the pool's computer system only if there is a sell

[6] Soft dollars are commission payments made by fund managers to brokerage firms, which are used partially, for example, to pay for research, analytics, and other services, enhancing investment decisions without direct cash payments.

order at a corresponding price. However, unlike on a *lit* market, orders that have not yet been executed are not visible to the traders and computerized trading systems using the dark pool.

Despite advancements, many fund managers still rely on investment banks for trading due to their influence and existing relationships. Commissions paid by fund-management firms enable them to meet soft-dollar obligations, receiving research and services at no direct cost. Fund managers frequently utilize their execution algorithms to break down substantial orders into smaller child orders. These child orders are often sent to investment banks for execution, with the bank's systems determining the order routing. Investment banks typically prioritize their own dark pools for order execution; if not feasible, orders may move to other banks' dark pools or public lit markets. Second-generation dark pools maintain the structural dependency of fund-management firms on investment banks while significantly increasing trading volumes compared to the first-generation. That brought a substantial gain of market share for dark pools, which increased from just over 4% of US share trading in 2008 to 17% by December 2014.

2007, 2018: In Europe, dark pools faced more skepticism, especially second-generation investment-bank-owned pools, leading to regulatory restrictions. In 2007, the Markets in Financial Instruments Directive (MiFID) implementation put in force new regulatory standards, affecting the landscape of dark pool trading in European markets. In 2018 MiFID II was introduced with the aim to encourage trading in lit venues or large-scale deals in first-generation dark pools by limiting the proportion of trading conducted in second-generation pools. This regulation also focused on reducing structural dependency by explicitly requiring fund managers to pay for research and other services from investment banks. The impact of MiFID II on reducing structural dependency is yet to be fully realized, but early signs suggest a potential decrease in dependence on investment banks for certain services.

2005–2018: For more than a decade, the SEC proposed and refined a new set of regulations on dark pools, with the final version being enacted in 2018.

2017: A critical milestone was reached when the volume of stocks exchanged within dark pools exceeded that of the NYSE for the first time.

Currently, over 60 dark pools are in operation, serving as alternative trading systems and catering to institutional investors seeking discretion and reduced market impact in their trades.

3.4 Discussions on Dark Pools

In the ontology of financial markets, dark pools exist as shadow entities, circumventing the luminosity of regulated marketplaces and altering the fundamental nature of trading interactions. They serve as discrete realms where entities can engage in financial exchanges obscured from the conventional transparency of mainstream markets. In this secluded space, they manifest as essential components, reshaping the intrinsic dynamics of market interactions and the essence of trade execution. Dark pools have emerged and developed predominantly due to regulatory fragmentation, technological advancements, and modifications in market structures.

To begin, the need for regulatory fragmentation was one of the catalysts behind the evolution of dark pools. Traditional trading platforms maintained a near-monopoly, prompting regulators to introduce fragmentation in capital markets. This fragmentation spurred the development of new platforms, like dark pools, and facilitated the comprehensive delivery of more adapted, cost-effective services to fulfill investor and trader needs. In response to the demand for greater transaction transparency, regulators structured systems of exemptions to the principles of pre-trade transparency, which was particularly beneficial for investors dealing with large orders, thus paving the way for the prominence of dark pools. However, as dark pools evolved into profitable ventures, there was a push for expansion. The need for more traders to grow has led some of these venues to paradoxically invite high-frequency traders—the very entities they were designed to evade. Nowadays, certain dark pools permit HFT and hence deviate from their original intention.

Technological advancements have played a crucial role in the development of dark pools. Improved communication networks, enhanced data processing speeds, and increased storage capacities have made these new platforms more efficient and reliable. Due to these technological innovations, the advent of HFT has further contributed to the proliferation of dark pools, serving, for instance, as possible liquidity providers. Recently, this aspect has been questioned and analyzed more in detail from theoretical and empirical perspectives, finding that the effect of high-frequency traders in dark pools is not as advantageous as one may think. A study by Aquilina et al. (2024) has shown, for instance, that HFTs rarely provide liquidity in the dark, but they rather consume liquidity, in particular in order to take advantage of stale reference prices, which play a significant role in increasing price impact. Moreover, illicit behaviors from HFT have been largely debated in the last few years. The case of the Attorney General of the State of New York against Barclays Capital Inc. and Barclays plc., is a notable example on this aspect, discussed in Clarke (2014). The reduction in tick size by US and European regulators in the 2000s has significantly impacted market structures, weakening market depth and reducing the capacity to absorb large orders without affecting prices. This alteration has proven beneficial for dark pools, attracting investors eager to place large orders without negatively impacting the market.

However, the demand for dark pools in Europe in recent years has increased partly due to regulatory changes regarding transparency. The MiFID, which came into force in 2004 and was implemented in 2007, set uniform pre-trade transparency requirements across European equity trading venues, with a few exceptions (through special waivers). Additionally, the regulation has made more pre-trade transparency mandatory in most European equity trading venues (MiFID I). Coupled with the growing presence of algorithmic traders who use latency advantages to obtain and act on information about supply and demand in order books, those led investors to seek dark trading alternatives.

3.4.1 Impact of Dark Pools on Financial Markets

In this section, we discuss a few widely recognized general beliefs regarding the impact of dark pools on market dynamics. Although dark pools provide some advantages for large investors, they face several criticisms centered mainly around their inherent opacity and the subsequent effects on the broader market. However, for several points raised below, conflicting conclusions have been drawn in the academic literature based on theoretical and empirical studies (see, e.g., Zhu 2014; Hintz 2014; Adrian 2015; Degryse et al. 2015; Johnson 2016; Buti et al. 2022). We can summarize the main criticisms as follows.

Loss of Price Discovery Price discovery pertains to how a trade's final price is determined based on the way new information is acquired and incorporated (see, e.g., O'Hara 2003; Brunnermeier 2005). Dark pools operate away from public eyes, which critics argue undermines the price discovery process. Price discovery is crucial in financial markets as it helps determine a security's fair market value. By trading in the shadows, dark pools may hinder the process of determining prices in open markets since the trading activities and the prices at which trades occur are not disclosed to the public. We should, however, mention that this phenomenon is strictly linked to the problem of information acquisition. As stated in Brogaard and Pan (2022)

> ... one cannot understand whether dark pools improve or harm price discovery—which is of first-order importance for the allocation of risk and capital—without understanding how dark pools affect information acquisition.

For instance, Zhu (2014) suggests that when trading in the lit market and the dark pool, there is a tradeoff between transaction costs and execution time. Dark pools offer lower transaction costs compared to exchanges but cannot guarantee execution, or execution is delayed. This setup seems to push *informed traders*, i.e., those traders who can rely on proprietary information on the intrinsic value of a traded asset, to send their orders to traditional markets so as not to be exposed to the risk of execution, and uninformed ones, also called liquidity orders, to dark pools (see also European Central Bank (2015, Section 2)). Some other contributions, for example, Nimalendran and Ray (2014), suggest that informed agents take advantage of concurrent trading in both venues. In particular the lit market is used for price discovery, and the dark pool to reduce costs and maximize profits.

The literature has proposed many models to explain the behavior of traders who have simultaneous access to the lit market and the dark pool, sometimes leading to conflicting results, which in fact, demonstrate that the price discovery mechanism is way more complex in reality. An overview of the existing theoretical and empirical results, and the corresponding underlying assumptions, are summarised in Brolley (2020, Tables 1 and 2), and for more references we also suggest Hendershott and Mendelson (2000), Zhu (2014), Buti et al. (2017), Degryse et al. (2009), Ye (2011), and Egginton et al. (2023) for the theoretical studies, and Degryse et al. (2015),

Nimalendran and Ray (2014), Comerton-Forde and Putniņš (2015), and Foley and Putniņš (2016) for empirical ones.

Fragmentation of Information and Liquidity One can also argue that dark pools contribute to market fragmentation. Unlike public exchanges, where all buy and sell orders are collected together to determine price, dark pools siphon off a portion of these orders, thus fragmenting the information and liquidity necessary for a unified market price. This fragmentation can lead to a reduction in the volume of transactions in the lit market, potentially resulting in higher spreads and price volatility.

Lack of Access Dark pools are often restricted to institutional or well-heeled investors, potentially creating a two-tiered market system. This lack of access for all market participants can lead to a fairness issue, as only some have the same opportunities or information. This can give rise to an unfair advantage for specific traders who have access to dark pools over those who do not.

Information Leakage Despite their private nature, there is a risk of information leakage of standing orders or the liquidity state of the pool. This risk often arises due to information asymmetry or fishing behaviors. Information asymmetry refers to the fact that some or all details about an order sent to the pool are made available to another trader (or to a liquidity partner), who can make use of this informational superiority to trade against the order. Fishing is the mechanism for deducing the presence and the size of an order by submitting frequent and small orders to the dark pool (that is, pinging the book) and using such information to manipulate the price in the lit market. We refer to Mittal (2008) for a thorough discussion on how information leaks and how it affects order execution. Since dark pools are supposed to be anonymous, any trade intention or other information leakage to other market participants can be detrimental.[7] Information leakage has been a concern for regulators, and legal actions have been conducted by the authorities. A few concrete examples are discussed below in Sect. 3.4.2.

Negative Externality on Traditional Lit Exchanges Dark pools introduce negative externalities[8] that can harm market integrity and investor fairness. As already

[7] A dark pool keeps order information until after a trade is executed. As argued in Mittal (2008), when this happens, certain market signals may still be generated, suggesting, for instance, the potential presence of further liquidity in the dark pool. This type of inference is perceived as innocuous since order sizes remain undisclosed and do not contribute to the phenomenon of information leakage.

[8] Roughly speaking, the term externality refers to the set of effects (either positive or negative) that an economic activity has on a third party without her/him being involved in such activity. Using the definition of Pigou, externalities arise when

> ... one person A, in the course of rendering some service for which payment is made to a second person B, incidentally, also renders services or disservices to other persons C, D and E

discussed in paragraph *Loss of price discovery*, informed traders possessing superior information about asset values (e.g., through proprietary data or advanced analysis) tend to concentrate on the lit market, where they can strategically capitalize on information asymmetries to trade against less-informed market participants, potentially exploiting their relative naivete. This practice is called cream skimming (see, for instance, Easley et al. 1996 for more explanation on the practices for incentivizing cream skimming of certain market participants). Cream skimming contributes to market fragmentation as trading volume migrates from centralized lit markets to alternative venues (O'Hara & Ye 2011). Consequently, traders in lit markets face increased costs due to reduced liquidity and wider bid-ask spreads (Hatheway et al. 2017). Paradoxically, while informed traders initially benefit from this dynamic, excessive cream skimming can lead to adverse selection, where they primarily trade against other well-informed players, diluting their informational advantage.

Lack of transparency in dark pools might also harm price discovery and accuracy and increase volatility (Ye 2011; Brolley 2020). In this sense, dark pools as alternative trading venues create a negative externality on the lit market. These issues also highlight the ongoing debate and regulatory effort to find the right balance between mitigating dark pools' harms and leveraging their potential benefits.

Impact on Volatility Petrescu et al. (2017) discuss how dark pool activities could either increase market volatility and instability or, conversely, enhance market stability by reducing volatility. This dichotomy aligns with earlier research; O'Hara and Ye (2011) and Garvey et al. (2016) suggest that moving orders from public exchanges to dark pools diminishes the information available in public order books, potentially increasing price volatility due to less information being available for price formation. On the other hand, Ye (2011), Buti et al. (2022), and Foley et al. (2012) provide a basis for the argument that dark pools might actually stabilize markets.

Overall, the arguments that dark pools might exacerbate market volatility may depend on the following mechanisms. Firstly, by running a portion of the trading activity, dark pools might cause a delay in price adjustments on public exchanges. This can lead to sudden and significant price movements once the information from dark pools is eventually incorporated into public market prices, thus potentially contributing to increased market volatility. Secondly, the absence of price transparency can exacerbate market turbulence in several ways. To begin with, it creates challenges in accurately assessing liquidity levels, which can lead to uncertainty and instability in the market. Moreover, delayed reactions to news can occur due to the lack of clear pricing information, generating additional market uncertainties

---

The problem of externality is a key issue in the economic literature that keeps attracting the attention of economists. We refer to Pigou (1924), Scitovsky (1954), Coase (1960), Buchanan and Stubblebine (1962), Dahlman (1979), and Hillman (2019) for more extensive studies on this topic.

and potentially amplifying volatility. Also, the limited transparency may encourage herding behavior among investors, as the absence of clear price signals can lead to a reliance on the actions of others rather than independent analysis. As a consequence, the market movements may be amplified (see Bikhchandani and Sharma (2000) and references therein for the causes of herding and its effects on financial markets). Moreover, the opacity in pricing can create opportunities for market manipulation by certain participants who possess privileged information, and this, in turn, can contribute to market volatility as highlighted in Aggarwal and Wu (2006).

Counter-Party Risk Engaging in dark pool trading involves interacting with multiple counterparties, such as other traders and liquidity providers, exposing traders to the inherent risk of default or counterparty failure. Koehn and Koehn (2014) emphasize that in the context of over the counter (OTC) dark pools with anonymized trades, investors face challenges as they cannot independently assess the identity of counterparties or evaluate the potential risks associated with their trades. Unlike transactions on public exchanges backed by clearinghouse guarantees, dark trades lack this protection against counterparty risk, making it difficult for investors to gauge the extent of risk involved.

Regulatory Concerns Dark pools often raise regulatory concerns given their opacity. Regulators worry about the potential for illicit activities, such as insider trading, that could go undetected in the dark pools due to the lack of transparency. Moreover, the regulatory framework for dark pools is often seen as less stringent than public exchanges, which might invite malicious practices or unfair trading advantages.

These criticisms underscore the tension between the private benefits (to institutional investors/or users of dark pools) that dark pools provide to their participants and the potential negative externalities they may impose on the broader market ecosystem. However, while dark pools have been criticized for their lack of transparency and potential for information and liquidity fragmentation, they also offer several notable benefits that serve as counterpoints to these criticisms.

Reduced Market Impact One of the primary advantages of dark pools is the reduced market impact of large trades. By allowing institutional investors to execute large orders away from the public eye, dark pools help to prevent significant price changes that would otherwise occur on public exchanges. This is especially beneficial for investors who manage large portfolios and require minimal market impact to preserve their execution price.

Increased Liquidity By providing a venue for large block trades, dark pools add to the overall liquidity of the financial markets. While they might fragment liquidity from public exchanges, they create a separate pool of liquidity that can benefit the market participants who have access to them. Such additional liquidity can be valuable during volatile market conditions. For a recent empirical study on how lit and dark market fragmentation affects liquidity, we refer to Gresse (2017). Based on data from eight stock exchanges, the author confirmed that liquidity improves with, or at worst, is not affected by, trading on several platforms.

Improved Execution Quality and Cost Efficiency Dark pools often improve execution quality by enabling trades to be executed at a price more favorable than the prevailing execution price in the lit market.[9] Such a mechanism is also documented in Koehn and Koehn (2014), mentioning that dark pools may assist investors in saving money by executing trades at, for instance, the midpoint between the best buying and selling prices, thus reducing extra costs such as bid-offer spreads and exchange fees. Hatheway et al. (2017) provide a thorough empirical analysis aimed at quantifying the correlation between dark trading and transaction costs. There is also evidence suggesting that dark pools may have contributed to the reduction of commission costs for institutional investors (McDermott 2011).

Investors frequently shift their trades away from lit exchanges, citing lower transaction costs as a motivating factor. However, the consistency of these lower costs remains undetermined. In Koehn and Koehn (2014) it is highlighted that lower trading costs may not be beneficial per se, and it is crucial to consider implicit or hidden costs beyond price spreads. Even if explicit costs are lower, the reason for such differences may require examination. For instance, lower transaction costs might be a result of reduced regulatory expenses. In Koehn and Koehn (2014), the authors also argue that the difference in regulatory costs not only gives dark networks a competitive advantage over public exchanges but also places investors at risk.

Privacy and Anonymity Dark pools offer an invaluable service for investors who prioritize privacy. The anonymity provided by dark pools allows investors to protect their trading strategies and positions from the broader market, which can be crucial for achieving their investment objectives.

Market Competition Dark pools introduce an element of competition to traditional exchanges, which can foster innovation and potentially lead to improved market structures and trading technologies. This competition can also drive down trading costs and improve services across the entire market ecosystem.

Flexible Trading Environment Dark pools often provide a more flexible trading environment than traditional exchanges. They allow for negotiated trades, which can be beneficial when the standard trading mechanisms of public exchanges are not ideal.

---

[9] For example, in US, there is the "trade-at rule" or "trade-at prohibition" associated with the Tick Size Pilot Program implemented by the U.S. SEC and the FINRA in 2016. In this context, the trade-at prohibition means that orders within certain pilot securities executed in dark pools must be at prices better than the displayed National Best Bid and Offer. This rule aims to promote price discovery and transparency by ensuring that dark pool executions improve upon publicly available market prices.

3.4.2 Regulatory Landscape and Legal Actions

In terms of regulation, dark pools are subject to various rules and requirements aimed at ensuring market integrity, investor protection, and transparency. For example, in the US, dark pools are regulated by the SEC and are subject to rules that require them to disclose information about their operations, provide fair access to all market participants, and prevent conflicts of interest. Dark pools are also required to register with the SEC and comply with various reporting and record-keeping requirements.

In Europe, dark pools are regulated under the MiFID and are subject to similar requirements regarding transparency, investor protection, and market integrity. MiFID requires dark pools to disclose information about their operations, provide fair access to all market participants, and prevent conflicts of interest. Dark pools are also required to register with national regulators and comply with various reporting and record-keeping requirements.

Regulatory changes under MiFID II/MiFIR introduced new transparency requirements and restrictions on the use of specific trading strategies, which might significantly impact the operation and growth of dark pools in Europe. For example, MiFID II/MiFIR requires dark pools to provide more pre-trade transparency and disclose more information about their operations, including the type of market data used, order handling and trading execution processes, and any particular priorities for various participants. MiFID II/MiFIR has also introduced volume caps on dark pool trading in specific instruments to limit their growth in Europe.

The U.S. SEC has taken substantial actions against multiple dark pools and alternative trading systems for breaching federal securities laws in recent years. Here is an overview of these actions.

In October 2011, Pipeline Trading System LLC faced penalties of $1 million for deceiving dark pool customers regarding order matching processes. In October 2012, eBX LLC was fined $800,000 for failing to safeguard dark pool customers' confidential trading data. In June 2014, Liquidnet Inc. had to pay $2 million for not adequately protecting dark pool customers' confidential trading information. July 2014 witnessed Lavaflow Inc. (a Citigroup unit and ATS operator) facing a $5 million penalty for allowing an affiliate to access and use customers' confidential trading information. In January 2015, UBS Securities LLC was fined $14.4 million for not properly informing all customers about an order type that provided advantages to specific traders. In August 2015, ITG Inc. (dark pool operator) and Alternet Securities (affiliated broker-dealer) faced penalties totaling $20.3 million for operating a secret trading desk that accessed confidential data about dark pool customers.

The most recent SEC enforcement actions date January 2016. SEC revealed significant enforcement actions against Barclays Capital Inc. and Credit Suisse Securities (USA) LLC for violating federal securities laws in relation to their operation of alternative trading systems, particularly dark pools and Credit Suisse's Light

Pool. Both firms were found to have failed to treat subscriber order information confidentially and failed to disclose certain aspects of their operations to subscribers.

Barclays agreed to settle the charges by accepting wrongdoing and agreeing to pay a total of $70 million in penalties, with $35 million going to the SEC and the other $35 million to the New York Attorney General's office. The investigation found that Barclays misrepresented various aspects, including its efforts to police its dark pool and the data feeds it utilized. Notably, Barclays misled subscribers about data connection speeds and surveillance regarding toxic trading.

Most precisely, as reported in the SEC declaration to Press (see SEC 2016),

> Barclays did not adequately disclose that it sometimes overrode Liquidity Profiling by moving some subscribers from the most aggressive categories to the least aggressive. The result was that subscribers who elected to block trading against aggressive subscribers nonetheless continued to interact with them. Barclays, at times, misrepresented the type and number of market data feeds that it used to calculate the National Best Bid and Offer in LX. For example, Barclays represented that it utilized direct feeds from exchanges to deter latency arbitrage when, in fact, Barclays used a combination of direct data feeds and other, slower feeds in the dark pool.

Credit Suisse settled the charges by agreeing to pay a substantial penalty amounting to $84.3 million, distributed between the SEC, the New York Attorney General's office, disgorgement, and prejudgment interest. Credit Suisse was found to have engaged in misrepresentation and failed to comply with regulatory requirements. Expressly, its Crossfinder misrepresenting how they characterized subscriber order flow and executing 117 million illegal sub-penny orders. Additionally, Light Pool, associated with Credit Suisse, was penalized for misrepresenting the Alpha Scoring, which is its method of measuring trading and handling opportunistic traders.

Some enforcement actions, along with some regulatory and congressional concerns over dark pools, are discussed for instance, in Shorter and Miller (2014) and Baxter (2017).[10]

3.4.3 Ethical and Moral Perspectives on Dark Pools

The notion of fairness is at the core of the ethical debate surrounding dark pools. The intentional concealment of trade details in these private venues may create a significant discrepancy between institutional investors and small investors, who, most of the time, do not have access to dark pools. From an ethical perspective, the opacity of dark pool trades raises concerns about fairness and equity within the financial system. Is it morally acceptable for a selected group of market participants to access an opportunity deliberately withheld from others? The epistemic value of hidden information in dark pools is also questioned, as it enables a privileged

[10] Additional information can be found at the Press Release page of the SEC website at https://www.sec.gov/news/pressreleases.

few to operate with an advantage, perhaps due to leakage. This ethical dilemma underscores the tension between the efficient market hypothesis[11] and equitable access to knowledge. Dark pools also challenge the efficient market hypothesis, suggesting that there is value in hidden or non-public information. From an ethical standpoint, this raises concerns about wealth distribution. In Mizuta et al. (2015), results from a multi-agent simulation study suggest that the impact of dark pools on market efficiency is not always positive and depends on the overall dark pool usage rate. Moreover, in the empirical study of Hatheway et al. (2017), it is demonstrated that market fragmentation is not welfare-improving in most cases.

With the growing prevalence of dark pool transactions in the market, opportunistic traders who previously executed their strategies on lit exchanges are now replicating those approaches within dark pools (see Johnson 2017). The opaque nature of these trading venues not only conceals such behaviors from regulators but, more disconcertingly, shields them from detection by other participants in the dark pool. Recent regulatory actions, media coverage, and the book of Lewis (2014) "Flash Boys" highlight fairness concerns. Allegations suggest that dark pool operators may have enabled high-frequency traders to front-run large institutional investors for a fee, misrepresenting the true nature of HFT in these alternative markets. There is also an increasing concern about manipulation, with some suggesting HFT firms strategically submit orders in the lit markets to deliberately influence securities prices within dark pools (Johnson 2017). Furthermore, persistent profits gained by certain participants through information leakage from dark pools could worsen income inequality.

Those may lead to some additional ethical and moral questions, such as: Does the pursuit of profit in dark pools come at the expense of broader social welfare? Should we be concerned about a financial system that appears to favor the wealthy and well-connected? In Cooper et al. (2020), three criteria are proposed to ensure the ethical operation of automated trading systems. One of the criteria is that the trading system should not circumvent transparent price discovery; hence, using dark pools or hidden orders cannot be seen as an effective financial market.

The reliability of dark pool trading and the perceptions of potential investors regarding this trading method may also represent significant concerns. While both reliability and perception are complex issues influenced by various factors, Koehn and Koehn (2014) reports that a substantial majority, about 67%, of long-term investors express difficulties trusting dark pools.

Finally, from a moral point of view, government regulation related to dark pools or, more generally, alternative trading venues also gives rise to questions about the role of government in safeguarding the interests of society. Is it morally justifiable for regulators to intervene in the operations of dark pools to ensure fairness and transparency, even if it means limiting individual freedoms and privacy? The ethical dilemma centers on the trade-off between personal liberties and the collective good, reflecting broader societal values.

---

[11] The efficient-market hypothesis posits that asset prices incorporate and reflect all the information that is currently accessible (Fama 1970).

References

Adrian, J. (2015). Informational inequality: How high frequency traders use premier access to information to prey on institutional investors. *Duke Law & Technology Review, 14*, 256.

Aggarwal, R. K., & Wu, G. (2006). Stock market manipulations. *The Journal of Business, 79*(4), 1915–1953.

Almgren, R., & Chriss, N. (2001). Optimal execution of portfolio transactions. *Journal of Risk, 3*, 5–40.

Aquilina, M., Foley, S., O'Neill, P., & Ruf, T. (2024). Sharks in the dark: Quantifying hft dark pool latency arbitrage. *Journal of Economic Dynamics and Control, 158*, 104786.

Baldacci, B., Manziuk, I., Mastrolia, T., & Rosenbaum, M. (2023). Market making and incentives design in the presence of a dark pool: A stackelberg actor–critic approach. *Operations Research, 71*(2), 727–749.

Baxter, B. P. (2017). The securities black market: Dark pool trading and the need for a more expensive regulation ATS-N. *Vanderbilt Law Review, 70*, 311.

Bayona, A. (2020, July). Dark pools and high frequency trading: A brief note. *Nota Técnica, 48*, B 21662–2012. Barcelona: IEF–Observatorio de Divulgación Financiera.

Bertsimas, D., & Lo, A. W. (1998). Optimal control of execution costs. *Journal of Financial Markets, 1*(1), 1–50.

Bikhchandani, S., & Sharma, S. (2000). Herd behavior in financial markets. *IMF Staff Papers, 47*(3), 279–310.

Bouchaud, J. P (2010). Price impact. In R. Cont (Ed.), *Encyclopedia of quantitative finance*. Hoboken: John Wiley & Sons.

Brogaard, J., & Pan, J. (2022). Dark pool trading and information acquisition. *The Review of Financial Studies, 35*(5), 2625–2666.

Brolley, M. (2020). Price improvement and execution risk in lit and dark markets. *Management Science, 66*(2), 863–886.

Brugler, J. (2015). Into the light: Dark pool trading and intraday market quality on the primary exchange. *Bank of England Working Paper* No. 545, available at SSRN: http://ssrn.com/abstract=2554621

Brunnermeier, M. K. (2005). Information leakage and market efficiency. *The Review of Financial Studies, 18*(2), 417–457.

Brunnermeier, M. K., & Pedersen, L. H. (2005). Predatory trading. *The Journal of Finance, 60*(4), 1825–1863.

Buchanan, J. M., & Stubblebine, W. C. (1962). Externality. In *Classic papers in natural resource economics* (pp. 138–154). Palgrave Macmillan.

Buti, S., Rindi, B., & Werner, I. M. (2017). Dark pool trading strategies, market quality and welfare. *Journal of Financial Economics, 124*(2), 244–265.

Buti, S., Rindi, B., & Werner, I. M. (2022). Diving into dark pools. *Financial Management, 51*(4), 961–994.

Cartea, Á., Jaimungal, S., & Penalva, J. (2015). *Algorithmic and high-frequency trading*. Cambridge University Press.

Clarke, T. (2014). High-frequency trading and dark pools: Sharks never sleep. *Law and Financial Markets Review, 8*(4), 342–351.

Coase, R. H. (1960). The problem of social cost. *The Journal of Law and Economics, 3*, 1–44.

Colaneri, K., & De Angelis, T. (2022). A class of recursive optimal stopping problems with applications to stock trading. *Mathematics of Operations Research, 47*(3), 1833–1861.

Comerton-Forde, C., & Putniņš, T. J. (2015). Dark trading and price discovery. *Journal of Financial Economics, 118*(1), 70–92.

Cooper, R., Davis, M., Kumiega, A., & Van Vliet, B. (2020). Ethics for automated financial markets. In *Handbook on ethics in finance* (pp. 1–18). Springer.

Cooper, R., Davis, M., & Van Vliet, B. (2016). The mysterious ethics of high-frequency trading. *Business Ethics Quarterly, 26*(1): 1–22.

Crisafi, M. A., & Macrina, A. (2016). Simultaneous trading in lit and dark pools. *International Journal of Theoretical and Applied Finance, 19*(08), 1650055.

Dahlman, C. J. (1979). The problem of externality. *The Journal of Law and Economics, 22*(1), 141–162.

Degryse, H., De Jong, F., & Kervel, V. V. (2015). The impact of dark trading and visible fragmentation on market quality. *Review of Finance, 19*(4), 1587–1622.

Degryse, H., Van Achter, M., & Wuyts, G. (2009). Dynamic order submission strategies with competition between a dealer market and a crossing network. *Journal of Financial Economics, 91*(3), 319–338.

Easley, D., Kiefer, N. M., & O'hara, M. (1996). Cream-skimming or profit-sharing? The curious role of purchased order flow. *The Journal of Finance, 51*(3), 811–833.

Egginton, J. F., McBrayer, G. A., & Watson, E. D. (2023). Shades of trade: Dark trading and price efficiency. *Journal of Banking & Finance, 155*, 107004 .

ESMA. (2020). ESMA report on trends, risks and vulnerabilities, N.2. https://www.esma.europa.eu/sites/default/files/library/esma_50-165-1287_report_on_trends_risks_and_vulnerabilities_no.2_2020.pdf

European Central Bank. (2015). Financial stability review–November 2015. https://www.ecb.europa.eu/pub/pdf/fsr/financialstabilityreview201511.en.pdf

Fama, E. F. (1970). Efficient capital markets: A review of theory and empirical work. *The Journal of Finance, 25*(2), 383–417.

Foley, S., Malinova, K., & Park, A. (2012). Dark trading on public exchanges. Available at SSRN 2182839.

Foley, S., & Putniņš, T. J. (2016). Should we be afraid of the dark? Dark trading and market quality. *Journal of Financial Economics, 122*(3), 456–481.

Gao, X., Zhou, X., & Zhu, L. (2018). Transform analysis for hawkes processes with applications in dark pool trading. *Quantitative Finance, 18*(2), 265–282.

Garvey, R., Huang, T., & Wu, F. (2016). Why do traders choose dark markets? *Journal of Banking & Finance, 68*, 12–28.

Gresse, C. (2017). Effects of lit and dark market fragmentation on liquidity. *Journal of Financial Markets, 35*, 1–20.

Hatch, R. (2009). Reforming the murky depths of wall street: Putting the spotlight on the security and exchange commission's regulatory proposal concerning dark pools of liquidity. *The George Washington Law Review, 78*, 1032.

Hatheway, F., Kwan, A., & Zheng, H. (2017). An empirical analysis of market segmentation on us equity markets. *Journal of Financial and Quantitative Analysis, 52*(6), 2399–2427.

Hendershott, T., & Mendelson, H. (2000) Crossing networks and dealer markets: Competition and performance. *The Journal of Finance, 55*(5), 2071–2115.

Hillman, A. L. (2019). *Public finance and public policy: A political economy perspective on the responsibilities and limitations of government*. Cambridge University Press.

Hintz, A. (2014). Lost in the dark: An analysis of the sec's regulatory response to dark pools. *DePaul Business & Commercial Law Journal, 13*, 329.

Horst, U., & Naujokat, F. (2014). When to cross the spread? Trading in two-sided limit order books. *SIAM Journal on Financial Mathematics, 5*(1), 278–315.

Johnson, K. N. (2016). Regulating innovation: High frequency trading in dark pools.*The Journal of Corporation Law, 42*, 833.

Johnson, K. N. (2017). Regulating innovation: High frequency trading in dark pools. *The Journal of Corporation Law, 42*(4), 833–836.

Koehn, D., & Koehn, D. (2014). Ethical issues with dark liquidity and the ethics of possible remedies. Seven Pillars Institute. Retrieved May 28.

Kratz, P. (2014). An explicit solution of a nonlinear-quadratic constrained stochastic control problem with jumps: Optimal liquidation in dark pools with adverse selection. *Mathematics of Operations Research, 39*(4), 1198–1220.

Kratz, P., & Schöneborn, T. (2015). Portfolio liquidation in dark pools in continuous time. *Mathematical Finance, 25*(3), 496–544.

Lewis, M. (2014). *Flash boys: A wall street revolt.* WW Norton & Company.

MacKenzie, D. (2019). Market devices and structural dependency: The origins and development of dark pools. *Finance and Society, 5*(1), 1–19.

McDermott, M. (2011, December 18). Shadows fall over dark pool trading practices. *Financial Times.* http://www.ft.com/intl/cms/s/0/4a2000bc-24b3-11e1-ac4b00144feabdc0.html#axzz2sNmAIP1Z

Mittal, H. (2008). Are you playing in a toxic dark pool? A guide to preventing information leakage. *The Journal of Trading, 3*(3), 20–33.

Mizuta, T., Kosugi, S., Kusumoto, T., Matsumoto, W., & Izumi, K. (2015). Effects of dark pools on financial markets efficiency and price discovery function: An investigation by multi-agent simulations. *Evolutionary and Institutional Economics Review, 12*, 375–394.

Nimalendran, M., & Ray, S. (2014). Informational linkages between dark and lit trading venues. *Journal of Financial Markets, 17*, 230–261.

O'Hara, M. (2003). Presidential address: Liquidity and price discovery. *The Journal of Finance, 58*(4), 1335–1354.

O'Hara, M., & Ye, M. (2011). Is market fragmentation harming market quality? *Journal of Financial Economics, 100*(3), 459–474.

Petrescu, M., Wedow, M., & Lari, N. (2017). Do dark pools amplify volatility in times of stress? *Applied Economics Letters, 24*(1), 25–29.

Pigou, A. C. (1924). *The economics of welfare*. Palgrave Mcmillan.

Scitovsky, T. (1954). Two concepts of external economies. *Journal of Political Economy, 62*(2), 143–151.

SEC. (2016). Barclays, credit suisse charged with dark pool violation. Firms collectively paying more than $150 million to settle cases. Available on line: SEC Press Release 2016-16.

Shorter, G., & Miller, R. S. (2014). Dark pools in equity trading: Policy concerns and recent developments. Library of Congress, Congressional Research Service.

Ye, M. (2011). A glimpse into the dark: Price formation, transaction cost and market share of the crossing network. Transaction Cost and Market Share of the Crossing Network (June 9, 2011).

Zhu, H. (2014). Do dark pools harm price discovery? *The Review of Financial Studies, 27*(3), 747–789.

Chapter 4
A Critical Discussion on how Agent-Based Modeling Participates in a Potential Paradigm Shift in Economics

Annalisa Fabretti

Abstract Economies are complex systems in which many entities with various properties and roles interact continuously. The methodological approach started with Leon Walras and Vilfredo Pareto and continued by J.F. Muth, K. J. Arrow, G. Debreu, R. Lucas, and P. Samuelson has been, so far, the fundamental paradigm under which economists work. The traditional approach uses analytically tractable models with a representative perfectly rational agent. In this framework, mathematics has been the main ally of economics. At the same time, economics and finance are experiencing, in recent decades, a new approach that some commentators claim is an attempt at a paradigm shift: from a representative rational agent approach towards a behavioral agent-based approach in which heterogeneous agents are seen as boundedly rational. In this regard, the present paper aims to discuss whether and how Agent-Based Modeling (ABM) is participating in this potential paradigm shift. ABM has gained momentum in recent years thanks to technological advances and the availability of big data. ABM can help address the complexity of economic problems that cannot be solved analytically, allowing researchers to simulate social and relational processes. A critical but non-exhaustive discussion is also devoted to exploring to what extent economics could be on the way to experiencing a paradigmatic change.

Keywords Agent-based modeling · Computational economics · Bounded rationality · Economic methodology

---

A. Fabretti (✉)
Department of Economics and Finance, Tor Vergata University of Rome, Rome, Italy
e-mail: Annalisa.Fabretti@uniroma2.it

E. Ippoliti et al. (eds.), *Philosophy and Finance: Ten Open Questions*, Synthese Library 504, https://doi.org/10.1007/978-3-031-96510-4_4

4.1 Introduction

Economics is a field that applies the approaches, methods, and tools of natural sciences to study social phenomena. Economic and social systems are complex, and different disciplines may approach the study of complexity in different ways. Since the seventeenth century, economics has undergone a transformation towards becoming more like a natural science through the process of mathematization, which was inspired by enlightenment thinking and newtonianism, and began with the work of Leon Walras and Vilfredo Pareto (Ingrao & Israel, 1990). The techniques and methodologies introduced during this process have strengthened economic theories and the discipline but have sometimes reduced interpretive freedom and produced a departure from empirical observation (Sitabhra & Chakrabarti, 2012). Despite inconsistent empirical evidence, the approaches of Pareto and Walras have remained the mainstream paradigm in economic modeling for centuries. However, there is growing evidence to suggest that this approach needs to be revised, and a paradigm shift in economics may be on the horizon, as argued by scholars such as Hommes, 2006, Davis, 2007, D'Orazio, 2017, Laybourn-Langton & Jacobs, 2018. Econophysics, behavioral economics, and technological progress are driving forces behind this shift, and ABM, as a product of these lines of research, plays an active role in contributing to this shift. This paper aims to investigate how ABM can contribute to this paradigm shift by examining how it has been incorporated into economic literature, particularly in finance, and how it has contributed to expanding knowledge in the field.

Furthermore, the paper explores the assumptions that underlie economic modeling and how ABM challenges and expands these assumptions by incorporating more realistic human behavior and heterogeneity and allowing for a dialogue between microstructure and macro outcomes. Nowadays, we know that the neoclassical standard of a human mind that decides by holding and processing any available information coldly through an objective calculation must be overcome (Simon, 1955; Kahneman & Tversky, 1979). On the other hand, the representative agent approach assumes identical economic agents, all acting rationally, while empirical evidence and daily life show us a heterogenous world. Thus, ABM allows one to consider more realistic human behavior and heterogeneity in human decision (Hommes, 2006; Tesfatsion, 2006).

Overall, these features allow ABM to enter into those cracks that have opened up due to an inability to fully adhere to empirical evidence and manage crises with substantial social impacts, such as the recent financial crises.

The discussion will be organized into four sections: a section is devoted to retracing the history of the economic theory broadly developed through two directions: (1) the mathematization process experienced in the nineteenth century; (2) the relationship between economics and physics with a focus on econophysics. The aim of this section consists in retracing the process by which economics transformed into natural science and adopted the current scientific paradigm funded by mathematical analysis; and how critical issues emerged to crack the fundamentals of the neoclassical paradigm. In this history, the roots of understanding where we are now are the seeds to spread new developments. Related to this, how and which paradigmatic changes economics has experienced so far is a matter of big discussion,[1] and few words must be spent in this direction; for this reason, in the same section, a specific paragraph titled "Exploring the Possibility of a Paradigm Shift in Economics" is devoted to this topic. The second section is devoted to ABM: definition, applications, and contribution to economics and finance. The third section will focus on the main debate: how ABM contributes to a paradigm shift. And finally, the last section concludes.

4.2 Economics: Where we Are

This section is devoted to retracing the history of economics as a discipline. As claimed by Dolfsma & Welch, 2009, a greater awareness of the history of economics may be an excellent vehicle for a better understanding of the role of paradigms in current economic thinking.

The focus will be directed toward the mathematization process that the field has experienced since the scientific revolution, the recent developments brought by behavioral economics and econophysics, and finally, a discussion about a potential paradigm shift in economics.

4.2.1 Economics and Mathematics: A Brief History of the Mathematization Process

Can Economics be considered a natural science? While it primarily deals with social phenomena, its methodological evolution has led to the adoption and integration of mathematical techniques, sparking a debate on its classification as a hard or soft science (Farmer, 2013; Mankiw, 2006). Going over the history of economics and, in particular, its process of mathematization, in the following, we try to understand better where economics is now and how its knowledge acquisition process works.

The Newtonian scientific revolution inspired economists to endow their field with tools borrowed from physics. Physics had become more accessible to the human mind thanks to the development and application of the scientific method and mathematics; at that time, all human disciplines, including the social sciences, perceived these approaches as new frontiers for human knowledge. Following Ingrao & Israel, 1990, the history of the theory of general equilibrium provides an excellent

[1] Note that notwithstanding related to the topics of the present paper, a comprehensive discussion will require a specific dissertation, and it is out of the present work. The reader can refer to Davis, 2007 and Dolfsma & Welch, 2009 for a more specific discussion.

example of reconstructing this process. Indeed, it was born from the intention of using mathematics as a tool both for description and for formalization and finally for analysis of economic phenomena. The development of this theory was part of a broader process that had as its objective the transformation of economics into a potentially more exact science, i.e., an ambitious transition from the status of social science as it is, as the study of social phenomena, to a natural science, as a user of the same techniques and methods of the natural sciences. A long and tiring process that started from the foundation of the basic concepts and has gone through various phases involving both the formal and the conceptual levels. The project to describe the economic reality with procedures similar to those adopted in the natural sciences was probably born from Turgot (1727–1781) and the physiocratic school. In these contexts, the need arises to represent economic quantities quantitatively, albeit with approximation; this need will later constitute a significant push toward the mathematization of the social sciences. Nicolas de Condorcet (1743–1794), Achylle Nicholas Isnard (1749–1803), Nicolas Francois Canard (1750–1838) and Jules Dupuit (1804–1866), all mathematicians and engineers, were all precursors of that process that led to transform the current approach to the study of economic systems. From their contribution, we derive, among other things (1) the idea of general economic equilibrium, formulated through the bond of dependence of all markets in a state of equilibrium; (2) the association between the concept of equilibrium, an already new idea of itself, with the theme of individual interest and the search for general harmony between these interests; (3) the first attempts to address the problem of the measurability of utility. Then, Leon Walras (1834–1910) and his successor Vilfredo Pareto (1848–1923) are considered the founders of economics as a mathematical science.[2] With Walras, we can say that the political economy acquires the character of a physical-mathematical science. In his view, the market can be seen as a place where blind and impersonal forces meet; thus, the study of this system can be approached through the determinism of rational mechanics. These forces, interacting with each other as forces of attraction on bodies, determine the price in the same way that gravitation determines the motion of the stars. The theoretical framework thus constructed can be studied through theorems and demonstrations on a par with classical mechanical systems. Such a program had been on the agenda of many economists throughout the following century. In the 1940s, Paul Samuelson (Samuelson, 1947) tried, in particular, to turn economics into a natural science by focusing on developing theorems subjected to empirical verification. But time after time, economics took a turn in which modeling, strongly on a mathematical ground, with the stress of axiomatic foundations, became more important than fitting the empirical observations. The mathematization and the application of analytical methods typical of physics require an abstraction and an idealization of the problems to be described, which could fall

[2] Simultaneously, William Stanley Jevons (1835–1882) in Great Britain and Carl Menger (1840–1921) in Austria also helped to develop that transformation that some historians consider a paradigmatic break, which has taken the conventional name of "marginalist revolution".

into reductionism. In economics, this involves the introduction of theoretically valid hypotheses but with delicate implications in practice and comparison with empirical evidence; an example among many is the construction of ideal exchanges with perfect competition and markets with perfect features. In perfect markets, producers and consumers rationally choose their supply and demand by optimizing their profit or utility. In theory, the maximization objective pushes producers and consumers towards adjustments of their supply and demand function which are reflected in competitive and efficient markets, assuming, in analogy with physics, immediate and cost adjustments of supply/demand; these assumptions lead to ideal situations that do not reflect the real world and markets. These limits emerge, for example, critically in the study of production processes where the auction process does not apply to the market as the adjustments are not precisely simultaneous but slow and expensive.

In the 1950s and 1960s, the large increase of game theory contributed to complete definitely the process of describing human action, decision, motivation, and strategies in mathematical terms. This epistemological process led to stress on the individual rationality in any human action, also characterized by selfishness and greediness, shortly identified into the concept of homo oeconomicus.

In summary, the representative agent, rational expectation, introduced by Muth, 1961 and popularized by Lucas 1971, and the efficient market hypothesis became the dominating paradigm in economics and finance around the sixties and seventies and now represent the current mainstream approach, usually referred to as neoclassical.

4.2.2 Economics Theory and Physics: A Long Fruitful Relationship from Newton to Econophysics

The relationship between physics and economics, as well their mutual influence, has a long history that has been extensively studied (Sitabhra & Chakrabarti, 2012; Ingrao & Israel, 1990). Since its inception, economics has drawn inspiration from the evolution of physics and the concept of balance between forces. Likewise, the study of economic phenomena has contributed to the development of physics. One notable example is the Brownian motion, which was initially developed by Bachelier in his PhD thesis of 1900 and later taken up by Einstein in 1905. Physics has inspired Economics also in

> the obsession of economics with the concept of maximisation of individual utilities. This is easy to understand once we remember that classical physics of that time was principally based on energy minimisation principles, such as the Principle of Least Action (Feynman 1964). We now know that even systems whose energy function cannot be properly defined can nevertheless be rigorously analysed, for example, by using the techniques of non-linear dynamics. However, academic disciplines are often driven into certain paths constrained by the availability of investigative techniques, and economics has not been an exception (cit p47 Sitabhra & Chakrabarti, 2012)

The transformation process aimed to promote and develop economics as a natural science through a deep mathematization process inspired by mechanics. However, something is missing, indeed:

> In its desperate effort to become a rigorous science by adopting, among other things, the formal mathematical framework of game theory, mainstream economics has become concerned less with describing reality than with an idealised version of the world. However, reality refuses to bend to the desire for theoretical elegance that an economist demands from his/her model. Unlike the utility maximising agents so beloved economists, in our day-to-day life we rarely go through very complicated optimisation processes in an effort to calculate the best course of action. Even if we had access to complete in formation about all the options available (which is seldom the case), the complexity of the computational problem would over whelm our decision-making capabilities. Thus, most often we are satisfied with choices that seem "good enough" to us, rather than the best one under all possible circumstances. Moreover, our choices may also reflect non-economic factors such as moral values that are usually not taken into consideration in mainstream economics (cit p2 p1 Sitabhra & Chakrabarti, 2012)

In the same years, in parallel, new developments laid the foundation for the current changes. Physics started focusing on non-equilibrium statistical mechanics, the theory of collective phenomena, nonlinear dynamics, complex systems and chaos, and various alternatives useful for describing economic and social phenomena (Sinatra et al., 2015). In the seventies, also, experimental results (Tversky & Kahneman, 1973) showed that in decision problems under uncertainty, humans do not behave rationally in the sense of maximizing expected utility, but their behavior follows more likely heuristic rules with significant biases. Starting with Simon (1955), behavioral economics had already discussed the realism of the representative agent and its rationality. In the 1980s and 1990s, empirical evidence and statistical analysis finally confirmed the doubts. From the economist's side, such discrepancies were addressed by behavioral economics (Shiller & Akerlof, 2009; Hommes, 2021). Outside the economists' community, scholars, among mathematicians, physicians, and computer scientists, looked at markets as complex systems populated by many interacting entities and showing chaotic dynamics (David, 1991). Such use of techniques belonging to physics for the understanding of phenomena of an economic nature has led to the birth of a new branch of study which has taken the name of econophysics.[3] Starting in the 1990s, a growing number of physicists have begun to develop their gaze and knowledge toward understanding social and economic phenomena, so much so that this new discipline, even if initially perceived as an intrusion or an invasion of the field, today cannot any longer be ignored; and indeed it should be appreciated for the contribution it has been able to make. It is a matter of fact that econophysics is a rapidly growing field (Mantegna & Eugene Stanley, 1999; Sitabhra & Chakrabarti, 2012). Physicists have brought their expertise in statistical physics and nonlinear dynamics to study complex systems, as their ability to handle large amounts of data and mathematical knowledge that exceeds the median economist's skills. Someone considered physicists' interest in

[3] The name was attributed by Eugene Stanley in a conference held in Kolkata in 1995

economics as meddling; however, the novel approach has brought a new perspective and has contaminated the traditional way of doing economics. The failure of economists to predict or prevent the 2008 economic crisis was truly a watershed in this sense. Afterward, econophysics began to be recognized much more, as it can witness the fact that the term was introduced in the New Palgrave Dictionary of Economics in 2008.

The success of econophysics well emerges by Sitabhra & Chakrabarti, 2012 words:

> Modelling itself on mathematics, mainstream economics is primarily deductive and based on axiomatic foundations. Econophysics seeks to be inductive, to be an empirically founded science based on observations, with the tools of mathematics and logic used to identify and establish relations among these observations. Econophysics does not strive to reinterpret empirical data to conform to a theorist's expectations, but describes the mechanisms by which economic systems actually evolve over time (cit p1 Sitabhra & Chakrabarti, 2012)

In recent decades, the introduction of these new perspectives has led economics and finance to experience new approaches that, some commentators claim, attempt at paradigm shift: from a representative rational agent approach towards a behavioral agent-based approach in which heterogeneous agents are seen as boundedly rational. Behavioral economics and econophysics have been the driving forces behind this potential change in paradigm (Sen 2009, Sitabhra & Chakrabarti, 2012*)*. Several scientists have argued that a change is needed in the way we investigate economic phenomena; in this regard, the failure that the discipline has experienced in the incapacity to prevent the 2008 financial crisis constitutes a piece of ironic empirical evidence of such urgency. This failure was depicted as:

> a systemic failure of the discipline. The roots of this failure have been traced to the dogmatic adherence to deriving elegant theorems from "reasonable" axioms, with complete disregard to empirical data (cit p2 p1 Sitabhra & Chakrabarti, 2012).

Looking back at the history of economics and its fruitful relationship with mathematics and physics, econophysics should be recognized as a continuation of this relationships, as also herald contamination of economics, that with other approaches, such as behavioral finance, for example, can lead to a significant improvement of economics theories and economics knowledge. Indeed, the combination of classical economics with new fields of investigation of physics can only be profitable; complex networks and game theory are an example of how the intertwining of these two fields can bear fruit: networks are aware of the space and the geometry in which people interact, game theory is aware more about the decision making and the strategy implementation, the interplay between these two kinds of complexity can lead to understanding more social interactions, development, hierarchies, exchanges, structures (Szabo & Fath, 2007; Manshaei, 2013; Kirman, 2016).

4.2.3 *Exploring the Possibility of a Paradigm Shift in Economics*

Whether economics requires a paradigm shift is a topic of ongoing discussion in the literature. In the following, a non-exhaustive discussion is presented with the aim of listing some critical points without the ambition to give answers.

First, it is essential to note that the term "economic paradigm" can refer to both politico-economic and theoretical-analytic frameworks. For the purposes of this paper, we will focus primarily on the theoretical-analytic framework, though we acknowledge the interconnection of the two paradigms. The former encompasses political and economic objectives, while the latter focuses on the functioning of economies and societies. Since the narratives which justify the goals and the social/economic adopted policy are often based on the analytical framework, the interconnection of the two paradigms cannot be ignored; however, in terms of discussing how and where the theoretical framework is going, the politico-economic paradigm will be kept in the background.

Following Kuhn, a change in paradigm occurs when a critical mass of anomalies appears to contradict the paradigm and emerges a successful alternative theory that better explains the empirical evidence (Kuhn, 1962). Lakatos argued that changes arrive through research programs gathering increasing support until they can replace old research programs. Given the interconnection of the politico-economical paradigm and the theoretical-analytical one in economics, such dynamics could meet forces that contrast or encourage the shift. Indeed it is not a case that the urgency of a rethinking in economic theory always arises after a financial crisis or economic shocks, as it happens for the politico-economical paradigm too (Laybourn-Langton & Jacobs, 2018).

According to Davis, 2007 there have been three revolutions in economic methodology; starting from a status quo originating from John Stuart Mill, running through John Keynes and the Cambridge School, and on to the Neo-Austrians. Dolfsma & Welch, 2009 say that economics has experienced paradigm shifts over time, and this will continue since the paradigmatic development of knowledge is inescapable. Dolfsma & Welch, 2009 also claim that the study of history helps to understand the impact of cultural perspective on the development of ideas and that "paradigm shifts in economics are not like bolts from the sky" (cit p1087 Dolfsma & Welch, 2009). New paradigms usually grow up from old ones and the development of knowledge is not linear or cumulative, thus, it happens that events are more likely to break the "normality".

The 2007 financial crisis brought to light the failure of the neoclassical paradigm in addressing financial systems. Financial bubbles, crises, and empirical discrepancies have inspired scholars to attempt new ways to understand financial systems. Some scholars had already perceived this urgency while looking at the dot.com bubble (2001) and the financial crash of 1987. In this regard, finance appears to be the ground on which the neoclassical paradigm met more criticism and showed its failure. Financial bubbles, crises, and empirical discrepancies have inspired scholars to attempt new ways to address financial systems. The financial crisis of 2007 is just the last in a row that pushed economics to deal with a rethinking of the neoclassical view, which dominates the theoretical framework. The reflections stimulated by the financial crisis have led majors to discuss the paradigm and to advocate the help of less conventional methodology (Trichet, 2010). The debate around the financial crisis involved the ability of classical economic theories to predict and prevent the upcoming situation and the possible responsibility those theories had in causing the problem. Just a few before the crisis, mainstream economists and politicians were claiming that all economic problems were solved and we had complete control of the economic cycles (see Lucas, 2003 and British prime minister Gordon Brown's declaration in the budget statement of 2007 (Brown, 2007), for example). Nowadays, we are living in a period in which change is wished, while mainstream economics still holds its dominion.

In this respect, the 2013 Nobel for Economies can be seen as symptomatic of the current coexistence of the two approaches (the neoclassical paradigm and the unconventional vision) and the ongoing shift. On the one hand, Eugene Fama, known for the efficient market hypothesis, represents the conventional view; on the other hand, Robert Shiller, a prominent exponent of behavioral finance, represents a new approach able to fill the discrepancy between theories and empirical evidence. On one side, the efficient market hypothesis claims the inexistence of speculative bubbles (the prices always reflect all the available information). On the other hand, behavioral finance explains market dynamics, price formation, and bubble emergencies, combining the emotionality with the rationality of investors. The efficient market hypothesis states that agents are rational and prices are always fine (in a sense, they reflect all the available information and are thus given by the actual price of future dividends), but investors are not always rational! Herbert Simon was one of the first to criticize the concept of rationality and to introduce the concept of bounded rationality. On one side, the homo oeconomicus framework sees rational human decisions as the results of some optimization process that leads to maximizing a utility or minimizing some costs; on the other, the idea that human behavior follows few roads to solutions, called heuristics, and eliminates many other options irrationally. Tversky & Kahneman, 1973, 1979 confirm this view and show that humans could violate rationality. Human beings, according to the neoclassical paradigm, are entirely emptied of their being human; the rationality of homo oeconomicus does not take into account that any human being has his/her own biography, that he/she is a socialized person with an experience and a personality that matter in any decision he/she takes; even if this decision is claimed to be technical and although it is taken as objective this never is.

The failure of economic theory in preventing crises or explaining human irrationality should remind economists that economics is a social science, and ignoring complex psychological, human, and social aspects in favor of simplistic and cold mathematical simplifications can bring to loose complexity and reality.

4.3 ABM: What it Is and how it Works

ABM is different from something well-defined. There has yet to be a consensus on an AB model or how we define such a model. ABM is a way of modeling characterized by multiple and heterogeneous interacting entities, that we usually call agents. Under this approach, systems are modeled as collections of autonomous and interacting agents. AB-models are designed at a micro level that "translates" into a macro-outcome. Macro features emerge from the aggregation of single decisions. The model can be (sometimes only partially) analytically studied or numerically implemented, in most cases deriving outcomes and findings through simulations.

The first attempt to use digital computation to model individual entities goes back to the 1950s. They were known as micro simulations, and the idea was essentially our modern idea of computational analysis; however, the computational tools needed to exploit the potentiality of those ideas completely were lacking at that time. Schelling's segregation model (dated back to the 1960s) is one of the most famous models in this branch. Since the nineties, the number of papers explicitly identified as using ABM increased exponentially to equate those identified by game theory and to overpass those identified as experimental economics (Axtell & Doyne Farmer, 2022).

AB-models are computational models able to overpass analytical tractability constraints. ABM encompasses innovative and evolutionary processes since agents can evolve, change, be born, die, grow; in other words they allow for any evolutionary changes in the composition of the agent population.

> [A]gents encapsulation gives ABM a striking resemblance to real-world system, . . . omissis.. ABM enables the modeling of systems encompassing nested self-referential subsystems as commonly see in the real world (cit p 8 Borrill & Tesfatsion, 2010).

Innovation processes are not easily modeled in traditional parameterized equation-based models. At the same time, ABM uses controlled computer experiments to investigate dynamics, verify aggregate results, to observe large-scale effects deriving from micro-level interactions. In this respect, ABM computer simulations can work as culture-dish experimentation.

AB-models are also known as Individual-Based Models (commonly in biology) or Multi Agent System (in computer science) and are applied in many fields. In some environments, it is referred to as synthetic systems; thus, agents are like synthetic particles or entities that interact artificially according to some mechanism. Despite the different denominations, all these approaches have the idea of developing computational models in common in which a pool of entities with specific characteristics is modeled at the very beginning. Then they are autonomous to move and evolve according to the setting, even die if foreseen by the modeler. Just to give some examples of the wide application, they are applied in agricultural and environmental economics (Berger, 2001; Gerst et al., 2013), industrial organization, automated markets, financial markets (LeBaron, 2006), business and management, electricity markets (Nicolaisen et al., 2000), labor markets (Tesfatsion, 1998, 2001; Fagiolo et al., 2004; Neugart, 2008); in biology and medicine as in ecological systems, epidemiology, health care management (Maes, 1990); and more, in social science representing collective behavior, crowding, the evolution of cooperation and trust, transportation system, urban planning, social influence, voting, social network formation, information storage management, and diffusion (see Squazzoni, 2010 and Squazzoni et al., 2014 for a review and a complete discussion on the application in social sciences).

ABM overlaps, borders, and relates with many fields: networks, game theory, econophysics, complex system and nonlinear dynamics, and deterministic chaos.

Nowadays, the large quantity of available data has been an engine for developing ABM and would be a great source to enlarge the applications. Indeed, ABM methodology is advanced enough to exploit our increasing computational power and data resources fully (D'Orazio, 2017). Nowadays, data can come from consumers and users profiles available from the web; or from laboratory and field experiments: data coming from experimental economics can be precious to designing new models (in particular to set behavioral rules), understanding and validating heuristics (Assenza et al., 2014). On its side, ABM can contribute to enlarging experimental results by overcoming material constraints that experimental economics may encounter as bounded resources or unverifiable and uncontrollable factors (Fabretti et al., 2017). The coupling of these two approaches nowadays appears very promising.

4.3.1 ABM in Economics and Finance

Finance is one of the most active areas of application of ABM. ABM's success in economics and finance is due to a confluence of factors, among which we can list (i) advances in computing hardware and software that allow implementing more and more complex models (in terms of the number of agents simulations, behavioral rules elaboration, and so on); (ii) high-quality data availability (iii) parallel development of behavioral and experimental economics; (iv) the coupling with data coming from laboratory and field experiments.

In Finance, ABM is particularly successful because it contributed to reproducing empirical evidence that the rational expectation approach cannot address or it cannot explain. Moreover, ABM was successfully used in regulatory policy evaluation and design and in studying systemic risk. In economics, AB-models are also broadly applied in micro and macroeconomics, labor economics, game theory, political economics, and environmental economics. However, finance is the ground in which that paradigm that ABM could contribute to mutate found more gaps and cracks. For these reasons, in the following, we retrace synthetically the contribution that ABM has had in this field since it is functional in reconstructing the crisis of the current paradigm and the solutions that new approaches can help to find.

According to the efficient market hypothesis, returns are normal and identically distributed, and prices are unpredictable. In 1963, Benoit Mandelbrot observed that financial time series show a peculiarity in volatility, well-known as volatility clustering (Mandelbrot, 1963). He observed that large changes more likely follow large changes, and similarly, small changes tend to be followed by small changes. In other words, high-volatility events tend to cluster in time, and according to this fact, we can argue for some predictability in this sense. Volatility clustering is incompatible with rational expectations and the efficient markets hypothesis: if markets are efficient, price changes can be explained only by exogenous events; however, only a small part of large movements can be associated with new information arriving outside the market (Cutler et al., 1989).

Financial prices, as already discussed in the previous section, have aroused the interest of many physicists, giving life to econophysics, which observed properties similar to those observed in physical systems in which a large number of entities interact, such as scaling and self-similarity (Cont, 2001; Mantegna & Eugene Stanley, 1999). Volatility clustering, as self-similarity and fractal properties of financial data, has been largely studied empirically and also modeled analytically more adherently with the empirical facts (see Svetlozar et al., 2011 for a comprehensive view). But despite this, we must stress that

> ABMs provide insight into the causes of clustered volatility, both yielding conceptual insight and providing better practical tools for forecasting volatility." (cit Axtell & Doyne Farmer, 2022 p 22)

Indeed, ABM contributed to the study of volatility clustering, with a specific interest in understanding the origin and other features shown by data (such as power law, among others). The Santa Fe stock market model (Arthur et al., 1997) was one of the first AB-model to investigate the origins of volatility clustering. In this model, agents trade a stock or a bond with two sources of information: fundamentals (dividends paid by the stock) and the behavior of prices summarized by technical indicators. Similarly, Brock & Hommes, 1997; Brock & Cars, 1998 investigate an asset pricing model with heterogeneous beliefs. These models, if all traders were identical and rational, essentially reduce to Lucas, 1978 asset pricing model, and under the additional assumption that dividends are IID, the asset price dynamics would reduce to one constant price equal to the fundamental. In a heterogeneous agent financial market, the evolutionary dynamics may lead to persistent deviations from the fundamental price and highly irregular, chaotic asset price fluctuations.

In this model, agents are modeled as boundedly rational, giving more realistic dynamics that match the properties of price changes and volatility. Among many models in the literature, one seminal model proposed by Lux and Marchesi (1999) deserves to be illustrated in detail as a prototype of a family of model.[4] Lux and Marchesi (1999) consider two groups of traders with three possible investment strategies and the possibility to switch between strategies, a mechanism that connects the probability to switch to the market dynamics and the observation of other traders. Here two strategies coexist in the market: one classical and coherent with the efficient market hypothesis and one practical; to be more precise, they introduce in this model a classification, which now represents almost a standard:

[4] The description of a specific model is intended to highlight the novel contribution of this methodology.

"fundamentalist" strategy which consists, coherently with EMH, in choosing to buy (or to sell) if the current price is below (or above) the fundamental price, which could be the expectation of the discounted future earnings; noise traders, also called chartist, which follows prices tendency. In their model, fundamental values follow a random walk, and noise traders can be optimistic or pessimistic; the former believe in a rise in prices, the latter in decline. Another seminal contribution of this model consists in allowing agents to switch between strategies with an adaptive and imitative mechanism based on following the most profitable strategy. Traders can move from one group to another with a certain probability which reflects the opinion change induced by market dynamics (endogenous probability). Noise traders can switch between pessimistic and optimistic. A rising price and a large majority of optimism increase the switching probability from pessimistic to optimistic. The probability of switching is a weighted combination of the actual price trend and the proportion of pessimistic and optimistic individuals; it follows, in this way, a herding mechanism. Prices are driven by the imbalance between supply and demand. An excess demand increases the price, while an excess supply induces a decrease. The stationary state is given by a price following, on average, the fundamentals. Hence in the long term, the market results efficiently. Simulated prices, in line with real markets, show volatility clustering. The alternation of calm and turbulence proceeds in parallel with changes in the group. High volatility periods are coupled with a large proportion of noise traders.

Such an approach of seeing market participants rely on both technical and fundamental trading rules to determine their orders has proven to be quite successful, also in exploring the impact of heterogeneous interacting agents upon the market dynamics. By Lux & Marchesi, 1999 model we can derive some crucial innovations contributed by ABM: agents are heterogeneous, they can adopt different strategies, and they can also switch among them. In the related literature, many papers follow similar schemes; to cite just some of the critical contributions: Day & Huang, 1990, De Grauwe et al., 1993, Lux, 1995, Brock & Hommes, 1997, Chiarella, 1992 and Farmer & Joshi, 2002. Comprehensive reviews are in Hommes, 2006, LeBaron, 2006 and Axtell & Doyne Farmer, 2022.

Other applications of ABM include normative understanding as for example, in predicting consequences of regulatory changes (see decimalization of Nasdaq occurred in 2001, Axtell & Doyne Farmer, 2022) or the evaluation and policy design (see the case of taxation in Pellizzari & Westerhoff, 2009).

Since ABM allows accounting for fads, contagion, domino effects, and imitative behavior, it is a suitable methodology to tackle systemic risk. Indeed systemic risk is strictly related to the common market sentiment, and it is particularly sensitive to financial contagion and domino effect. The European Central Bank and Bank of England are starting to use ABM as alternative tools to evaluate risk at light of the recent financial crisis (see for example Trichet's speech during the Opening Address at the ECB Central Banking Conference 2010).

Another contribution of ABM stays in the comprehension of the square root law of market impact; under rational expectation, market impact is expected to be linear. ABM helps to show that this empirical law could be understood in terms of a combination of the market microstructure and the fact that traders exploit and absorb arbitrage opportunities (Bouchaud et al. 2015).

In conclusion, so far ABM in finance has reached a development stage such that they can reproduce qualitatively and quantitatively market features and provide forecasts, also thanks to a large literature involved in their calibration, which has been crucial for ABM's success and which has challenged scholars for years (Gilli & Winker, 2003; Alfarano et al., 2005; Bianchi et al., 2007; Windrum et al., 2007; Winker et al., 2007; Franke, 2009; Fabretti, 2013; Grazzini & Richiardi, 2015; Recchioni et al., 2015). The use of synthetic agents, coupled with artificial intelligence and potentially exploiting large amount of data constitutes a new frontier in many fields. This issue will be discussed more deeply in the next section.

4.4 Discussion

There is an open debate, but also consensus and general evidence that economics is experiencing changes in thinking, in research agenda, and in methodology (Davis, 2007; D'Orazio, 2017; Kirman, 2016; Hommes, 2006; Axtell & Doyne Farmer, 2022 among others). In this paper, we tried to tackle the contribution that ABM can have in this ongoing revolutionary transformation. A few questions arise: what is new in ABM and complex system approach? Is ABM just a numerical extension (or support) of mainstream economic theory? Is ABM complementary or a substitute for neoclassical economic theory?

Many features of ABM conspire to relax some of those strong assumptions on which relies the mainstream mathematical modelization. At the same time, more realistic human behavior can be incorporated into models. Computational economics and behavioral economics are replying better to the challenges communities, and social systems are facing right now. ABM, as part of this broad area of application, has met so far success but also skepticism. Their development can be compared to those of other areas of economics, such as game theory that, until the 1990s, almost 40 years after its birth, was confined to the mathematics departments; similarly, it occurred with behavioral and experimental economics, which began appearing in the late 1950s but accepted in mainstream economics department around 1990s (Axtell & Doyne Farmer, 2022). History repeats itself, thus no surprise to see a new field that meets skepticism or decisive resistance, and then from time to time, it conquers some space among researchers, as the number of papers published in this area can witness (Axtell & Doyne Farmer, 2022).

Many authors recognize the contribution that the computational approach has brought to many fields of social science as tools that can face social complexity (Squazzoni, 2010). The aggregate complexity derived by the interaction of many entities is hard to understand; in this, a micro-macro mapping can make the difference, as well focused by Squazzoni et al. (2014)

> even in ideal situations, it is hard to understand the aggregate behavior of groups of people whenever they interact over any significant period of time. This is due to the nonlinear interaction effects between individuals and the interplay between individual behavior and social dynamics and structures. These complications make any micro–macro mapping extremely difficult: Even quite comprehensive knowledge of behavior at a small scale does not necessarily help in understanding implications at a macro scale (e.g., Miller & Page, 2007). This is important as many social, economic, and political issues that have serious consequences on human well-being may pivot upon such complex group behavior. Examples of these consequences include economic crises, civil wars, and environmental problems, all of which occur on spatially large, real-time scales and take place within and between different social groups. All of these have macro implications, such as public budget restrictions and increasing urban ethnic segregation. These macro implications may have, in turn, important micro-level counterparts, such as increasing perceived insecurity and discriminative attitudes. Therefore, managing such large-scale problems requires a systematic understanding of the processes at both the individual and the group level—in other words, we cannot ignore the micro–macro mapping if we are to address these problems (cit. Squazzoni et al., 2014)

It is a matter of fact that ABM is an innovation that gained deserved momentum. On the one hand, we cannot deny that its success is linked to the computational revolution we are experiencing; very advanced technology, supercomputers, artificial intelligence, and the availability of large amounts of diversified, sophisticated, and detailed data can only be a push towards new horizons. In this sense, one could object that computational economics is just a complement of traditional and mainstream approaches; in other words, it could be a way of addressing numerically something that cannot be solved analytically. Is that the case? The skeptic could claim that once we allow for bounded rationality and heterogeneity, ABM just extends our knowledge by means of simulations. However the ABM contribution does not stop here and to completely address this new point we need to introduce another related issue: big data. Big data, without any doubt, represent a new frontier and a new opportunity for gaining human knowledge. The debate on the ontology and the existence of big data is wide; without pretending to be precise, we follow the definition that involves three main qualities: the magnitude of data, the complexity of the structure, and the speed at which data are collected (or produced) and stored, according to the available technologies, that nowadays make the difference. Despite their appeal, such amount and complexity of data require tools to manage them, and conventional and statistical methods could be inappropriate.

> In the case of big data-sets, conventional methods cannot be 'extended' or 'scaled up' in order to deal with the higher dimensionality and complexity. BD are indeed characterized by some peculiar features which need to be treated according to different techniques, such as: heterogeneity, noise accumulation, spurious correlation and incidental endogeneity. (cit. p413 D'Orazio, 2017)

Economists should invest in learning how such data can produce knowledge, and the discovery and adoption of new methodologies that best fit incumbent research needs cannot be postponed. D'Orazio (2017) deems that the interaction between big data and computational behavioral macroeconomics is contributing, in particular in macroeconomics, to the emergence of a new paradigm that is replacing the "systemic obsolescence of The Neoclassical_New Keynesian Synthesis approach". D'Orazio, 2017

> argued that the diffusion of BD in social sciences implies a deep epistemological change which will lead economists to consider (or re-consider)(a) the process of research, (b) the way in which they engage with data analysis, data processing and data gathering and (c) the constitution of knowledge. (cit p 423 D'Orazio, 2017)

Concluding, the availability of big data in addressing economic inquiries can lead to new styles of thinking and investigation. Still, it requires tools to tackle such kinds of data. ABM can integrate and be integrated with big data, since the latter provides micro information about agents and the former a way to aggregate such information to shed light on the macro observed picture. In this sense, ABM can be seen as a laboratory for social data, a way of looking at social and relational processes and reproducing observed dynamics. The in-vitro approach provided by ABM coupled with the availability of data is, in this sense, strongly powerful and promising. Thus we can affirm now that ABM is not just a computational extension of some relaxation of classical economic theory, but rather a flexible and powerful methodology able to face new economic challenges.

4.5 Conclusion

In conclusion, previous discussions have tried to reconstruct the picture of the contribution that ABM has brought to economics and, in particular, in finance. Janssen et al., 2008 say that ABM is

> ..an important new tool for modelling the dynamics of complex systems... (omissis) ... agent based modeling represents a new research methodology and a new way of conceptualizing research problems (cit p1 Janssen et al., 2008)

Moreover Hommes, 2006 says

> Economics and finance are witnessing an important paradigm shift, from a representative, rational agent approach towards a behavioral, agent-based approach in which markets are populated with boundedly rational, heterogeneous agents using rule of thumb strategies. (Hommes, 2006).

Another voice says:

> Unfettered by the mathematical strictures that constraints economic models that must be written down and solved analytically, ABMs serve as a potent new methodology for accelerating progress in economics, (cit. p 6 Axtell & Doyne Farmer, 2022).

But this is not enough to claim without doubt that economics need a renovation, indeed:

> Computational economies composed of software agents represent a new paradigm for economic research. Their role in substantive model building is yet nascent with some clear successes, as in finance. Nonetheless, many areas remain relatively untouched by this new technique. (cit. p 49 Axtell & Doyne Farmer, 2022)

In conclusion, no doubts that we have a potent, innovative methodology to face economic problems. There is a general agreement that ABM is a new methodology able to address the complexity better than standard economic theory. However we register two tendencies: who believe that we are witnessing a new paradigm (Hommes, 2006) and those who believe that conventional methods must be revised in the light of new methodology and empirical results, but ABM and mainstream economics are not competing (Axtell & Doyne Farmer, 2022). In this sense, ABM is seen more as a complement to conventional economic modeling than a substitute. By another hand, the need for a renovation in economic thinking appears urgent and inescapable (Davis, 2007; D'Orazio, 2017; Dolfsma & Welch, 2009) and ABM has participated so far with success in such a rethinking.

It far to say that computational economics must replace mainstream economics; on the one hand, mainstream economics must accept to drop some classical assumptions and open to novelties as ABM; on the other hand, ABM practice by its side could be more rigorous and trustful whether any model would comprise at least a theoretical skeleton or an analytical companion to be studied traditionally. Indeed, simulations are fascinating, and since computational tools and computational power are widely available nowadays, there is a concrete risk of abuse of simulations with huge models. Indeed the use of simulations to overpass theoretical tractability burdens and obstacles is very charming; however, we must keep in mind that the temptation of having an all-embracing model can make us fall down in an uncontrolled reality, leading to not reproducible results.

References

Alfarano, S., Wagner, F., & Lux, T. (2005). Estimation of agent-based models: The case of an asymmetric herding model. *Computational Economics, 26*, 19–49.

Arthur, W. B., Holland, J. H., LeBaron, B., Palmer, R., & Tayler, P. (1997). Asset pricing under endogenous expectations in an artificial stock market. In W. B. Arthur, S. Durlauf, & D. Lane (Eds.), *The economy as an evolving complex system II* (pp. 15–44). Addison-Wesley.

Axtell, R. L., & Doyne Farmer, J. (2022). Agent-based modeling in economics and finance: Past, present, and future. *Journal of Economics Literature*. (forthcoming).

Assenza, T., Bao, T., Hommes, C. H., & Massaro, D. (2014). Experiments on expectations in macroeconomics and finance. In *Experiments in Macroeconomics in research in experimental economics, Vol 17* (pp. 11–70). Emerald Group Publishing Limited.

Berger, T. (2001). Agent-based spatial models applied to agriculture: A simulation tool for technology diffusion, resource use changes and policy analysis. *Agricultural Economics, 25*, 245–260.

Bianchi, C., Cirillo, P., Gallegati, M., & Vagliasindi, P. (2007). Validating and calibrating agent-based models: A case study. *Computational Economics, 30*, 245–264.

Borrill, P. L., & Tesfatsion, L. (2010). *Agent-based modeling: The right mathematics for the social sciences?* Iowa State University, Department of Economics.

Brock, W. A., & Hommes, C. H. (1997). A rational route to randomness. *Econometrica, 65*, 1059–1095.

Brock, W. A., & Cars, H. H. (1998). Heterogeneous beliefs and routes to chaos in a simple asset pricing model. *Journal of Economic Dynamics and Control, 22*, 1235–1274.

Brown, G. (2007). *Budget statement to the house of commons*. Accessed Mar 30, 2023, from https://publications.parliament.uk/pa/cm200607/cmhansrd/cm070321/debtext/70321-0004.htm
Chiarella, C. (1992). The dynamics of speculative behaviour. *Annals of Operations Research, 37*, 101–123.
Cont, R. (2001). Empirical properties of asset returns: Stylized facts and statistical issues. *Quantitative Finance, 1*, 223–236.
Cutler, D. M., Poterba, J. M., & Summers, L. (1989). What moves stock prices? *Journal of Portfolio Management, 15*, 4–12.
Day, R. H., & Huang, W. (1990). Bulls, bears and market sheep. *Journal of Economic Behavior and Organization, 14*, 299–329.
David, A. H. (1991). Chaos and nonlinear dynamics: Application to financial markets. *The Journal of Finance, 46*(5), 1839–1877.
Davis, J. B. (2007). The turn in economics and the turn in economic methodology. *Journal of Economic Methodology, 14*, 275–290.
De Grauwe, P., Dewachter, H., & Embrechts, M. (1993). *Exchange rate theory: Chaotic models of foreign exchange markets*. Blackwell.
D'Orazio, P. (2017). Big data and complexity: Is macroeconomics heading toward a new paradigm? *Journal of Economic Methodology, 24*, 410–429.
Dolfsma, W., & Welch, P. J. (2009). Paradigms and novelty in economics: The history of economic thought as a source of enlightenment. *American Journal of Economics and Sociology, 68*, 1085–1106.
Fabretti, A. (2013). On the problem of calibrating an agent based model for financial markets. *Journal of Economic Interaction and Coordination, 8*, 277–293.
Fabretti, A., Gärling, T., Herzel, S., & Holmen, M. (2017). Convex incentives in financial markets: An agent-based analysis. *Decisions Economics and Finance, 40*, 375–395.
Fagiolo, G., Dosi, G., & Gabriele, R. (2004). Matching, bargaining, and wage setting in an evolutionary model of labor market and output dynamics. *Advances in Complex Systems, 7*(2), 157–186.
Farmer, J. D. (2013). Hypotheses nonfingo: Problems with the scientific method in economics. *Journal of Economic Methodology, 20*, 377–385.
Farmer, J. D., & Joshi, S. (2002). The price dynamics of common trading strategies. *Journal of Economic Behavior and Organization, 49*, 149–171.
Franke, R. (2009). Applying the method of simulated moments to estimate a small agent-based asset pricing model. *Journal of Empirical Finance, 16*, 804–815.
Gilli, M., & Winker, P. (2003). A global optimization heuristic for estimating agent based models. *Computational and Statistical Data Analysis, 42*, 299–312.
Grazzini, J., & Richiardi, M. (2015). Estimation of ergodic agent-based models by simulated minimum distance. *Journal of Econonmic Dynamics and Control, 51*, 148–165.
Gerst, M. D., Wang, P., Roventini, A., & Fagiolo, G. (2013). Agent-based modeling of climate policy: An introduction to the ENGAGE multi-level model framework. *Environmental Modelling & Software, 44*, 62–75.
Hommes, Cars H. 2006. Heterogeneous Agent Models in Economics and Finance. In Handbook of Computational Economics vol 2, L. Tesfatsion, K.L. Judd, 1109–1186 Elsevier.
Hommes, C. (2021). Behavioral and experimental macroeconomics and policy analysis: A complex systems approach. *Journal of Economic Literature, 59*(1), 149–219.
Ingrao, B., & Israel, G. (1990). *The invisible hand: Economic equilibrium in the history of science.*, translated by Ian McGilvray. MIT Press.
Janssen, M. A., Alessa, L. N.'i., Barton, M., Bergin, S., & Lee, A. (2008). Towards a community framework for agent-based modelling. *Journal of Artificial Societies and Social Simulation, 11*(2), 6.
Kahneman, D., & Tversky, A. (1979). Prospect theory: An analysis of decision under risk. *Econometrica, 47*(2), 263–291.
Kirman, A. (2016). Networks: A paradigm shift for economics? In Y. Bramoullé, A. Galeotti, & W. Brian (Eds.), *The Oxford handbook of the economics of networks*. Rogers Oxford Handbooks.

Kuhn, T. S. (1962). *The structure of scientific revolutions* (1st ed.). University of Chicago Press.
Laybourn-Langton, L., & Jacobs, M. (2018). Paradigm shifts in economic theory and policy. *Intereconomics, 53*, 113–118.
LeBaron, Blake. 2006. Agent-based computational finance. In Handbook of computational economics, vol. 2: Agent-based computational economics, L. Tesfatsion, K. Judd 1187–1233. North-Holland.
Lucas, R. E. (1978). Asset prices in an exchange economy. *Econometrica, 46*, 1429–1445.
Lucas, R. E. (2003). Macroeconomic Priorities. *American Economic Review., 93*, 1–14.
Lux, T. (1995). Herd behaviour, bubbles and crashes. *Economic Journal, 105*, 881–896.
Lux, T., & Marchesi, M. (1999). Scaling and criticality in a stochastic multi-agent model of a financial market. *Nature, 397*, 498–500.
Maes, P. (1990). *Designing autonomous agents: Theory and practice from biology to engineering and Back* (Vol. 6, pp. 1–2). MIT Press.
Mandelbrot, B. (1963). The variation on certain speculative prices. *Journal of Business, 36*, 394–419.
Mankiw, N. G. (2006). The macroeconomist as scientist and engineer. *Journal of Economic Perspectives, 20*, 29–46.
Manshaei, M. H., Zhu, Q., Alpcan, T., Basar, T., & Hubaux, J. P. (2013). Game theory meets network security and privacy. *ACM Computing Surveys, 45*, 1–39.
Mantegna, R. N., & Eugene Stanley, H. (1999). *Introduction to econophysics: Correlations and complexity in finance*. Cambridge University Press.
Muth, J. F. (1961). Rational expectations and the theory of price movements. *Econometrica, 29*, 315–335.
Nicolaisen, J., Petrov, V., & Tesfatsion, L. (2000). *Market power and efficiency in a computational electricity market with discriminatory double-auction pricing*. In Staff General Research Papers. Iowa State University.
Neugart, M. (2008). Labor market policy evaluation with ACE. *Journal of Economic Behavior and Organization, 67*(2), 418–430.
Pellizzari, P., & Westerhoff, F. (2009). Some effects of transaction taxes under different microstructures. *Journal of Economic Behavior & Organization, 72*(3), 850–863.
Recchioni, M. C., Tedeschi, G., & Gallegati, M. (2015). A calibration procedure for analyzing stock price dynamics in an agent-based framework. *Journal of Econonmic Dynamics and Control, 60*, 1–25.
Samuelson, P. A. (1947). *Foundations of economic analysis*. Harvard University.
Shiller, R., & Akerlof, G. (2009). *Animal spirits: How human psychology drives the economy and why it matters for global capitalism*. Princeton University Press.
Simon, H. A. (1955). A behavioral model of rational choice. *The Quarterly Journal of Economics, 69*, 99–118.
Sinatra, R., Deville, P., Szell, M., Wang, D., & Barabasi, A. l. (2015). A century of physics. *Nature Physics, 11*, 791–796.
Sitabhra, S., & Chakrabarti, B. (2012). Econophysics: An emerging discipline. *Economic and Political Weekly, 47*, 44–65.
Squazzoni, F. (2010). The impact of agent-based models in the social sciences after 15 years of incursion. In *History of economic ideas 18*. Istituti editoriali e poligrafici internazionali. Fabrizio Serra 1724-2169.
Squazzoni, F., Jager, W., & Edmonds, B. (2014). Social simulation in the social sciences: A brief overview. *Social Science Computer Review, 32*, 279–294.
Svetlozar, T. R., Kim, Y. S., Bianchi, M. L., & Fabozzi, F. J. (2011). *Financial models with levy processes and volatility clustering*. John Wiley & Sons, Inc.
Szabo, G., & Fath, G. (2007). Evolutionary games on graphs. *Physics Reports, 446*, 97–216.
Tesfatsion, L. (1998). Ex ante capacity effects in evolutionary labor markets with adaptive search. In *Economic Report*. Iowa State University.
Tesfatsion, L. (2001). Structure, behavior, and market power in an evolutionary labor market with adaptive search. *Journal of Economic Dynamics and Control, 25*, 419–457.

Tesfatsion, L. (2006). Agent-based computational economics: A constructive approach to economic theory. In L. Tesfatsion & K. L. Judd (Eds.), *Handbook of computational economics* (Vol. 2, pp. 831–880). Elsevier.
Trichet, J. -C. (2010). *Reflections on the nature of monetary policy non-standard measures and finance theory, Speech during the Opening Address at the ECB Central Banking Conf.* Accessed Mar 20, 2023, from https://www.ecb.europa.eu/press/key/date/2010/html/sp101118.en.html
Tversky, A., & Kahneman, D. (1973). Availability: A heuristic for judging frequency and probability. *Cognitive Psychology, 5*, 207–232.
Windrum, P., Fagiolo, G., & Moneta, A. (2007). Empirical validation of agent-based models: Alternatives and prospects. *Journal of Artificial Societies and Social Simulations, 10*(2), 1–8.
Winker, P., Gilli, M., & Jeleskovic, V. (2007). An objective function for simulation based inference on exchange rate data. *Journal of Economic Interaction and Coordination, 2*, 125–145.

Chapter 5
Money on my Mind: An Investigation of the Folk Classification of Money

Francesco Guala, Alexander Noyes, and Frank C. Keil

Abstract Little is known about the folk ontology of money. Across four studies, we examined how participants classify two important cases: currency artifacts (bills and coins) and *ad hoc* currency objects, which include artifacts not intended to be currency (e.g. cigarettes) and natural objects (e.g. shells). In all studies, participants reliably categorized shells and cigarettes as money only when they functioned as a currency within the current social system. In contrast, many participants continued to classify currency artifacts as money even when they no longer functioned as currency. Interestingly, responses were bimodal, with participants either fully endorsing or rejecting the classification of defunct currency artifacts as money. When asked to generate properties of money, participants mentioned social (e.g. being a medium of exchange) and physical (e.g. having a serial number) properties, but rated social properties as more important. Participants saw artifacts and social conditions as interacting but placed more weight on social conditions overall. Therefore, data indicated that although participants primarily represented money as a social kind, the intuition that broken or unused artifacts retain their intended kind led people to sometimes classify as money those artifacts that exist in virtue of social conditions.

Keywords Categorization · Concepts · Causal reasoning · Social cognition · Money

F. Guala (✉)
Department of Philosophy, University of Milan, Milan, Italy
e-mail: francesco.guala@unimi.it

A. Noyes
Department of Psychology, New York University, New York, NY, USA

F. C. Keil
Department of Psychology, Yale University, New Haven, CT, USA

E. Ippoliti et al. (eds.), *Philosophy and Finance: Ten Open Questions*, Synthese Library 504, https://doi.org/10.1007/978-3-031-96510-4_5

5.1 Introduction

Economic activities like trading, investing, or saving involve the use of many objects with peculiar social powers: paper bills, credit cards, cheques, bank accounts, and contracts, to name just a few. In general, these objects are significant because they confer social powers to individuals who otherwise lack them, such as the right to use certain material and social assets. But they are noteworthy also because their powers have little to do with their physical structure, and much to do with social facts and mechanisms, such as the laws, policies, and reoccurring behaviors that involve these objects.

In spite of the important role they play in our lives, institutional entities like bills and bank accounts are still relatively obscure and under-researched from a psychological perspective. In this paper we seek to provide a fuller picture of how people represent *money*, as a case study into the nature of institutional representations. Although money is only one of many institutional kinds, it is a paradigmatic one, enjoying a special status in the philosophy of social institutions (e.g., Simmel, 1978; Searle, 1995; Guala, 2016; Smit et al., 2016; Frasser & Guzmán, 2020; Mäki, 2020): it is, in fact, one of the key institutional kinds that an account of social institutions is expected to address. This special status is well-earned because of money's ubiquity and social significance. Thus, any psychological account of institutions ought to provide an adequate account of how people represent money. As individuals interact with money within larger institutions like markets and states, how do they represent the kind they interact with?

5.1.1 *Philosophical Perspectives on Money*

Money is an institutional kind. Institutions are reoccurring, rule-governed patterns of social behavior, which remain stable by constraining and guiding individuals' actions. For example: we would find it challenging to enter a store in the United States and make purchases with objects other than U.S. currency, even if the objects were independently valuable. But institutions also make new forms of acting possible: we can widely exchange dollar bills for goods and services, even if we do not know the consumption preferences of the seller, and even though the pieces of paper are not independently valuable. This universal medium of exchange enables efficient, widescale, and coordinated economic behavior that would be hard to achieve through other means (e.g. barter).

The status of institutional kinds is controversial. A prominent approach, which is popular among scientists, is to identify these kinds with the functions they fulfil. For example: money is anything that plays the role of medium of exchange, store of value, unit of accounting. Some philosophers are willing to follow this functionalist approach (e.g. Rosenberg, 2016; Guala, 2016), but others argue that fulfilling the appropriate functions is not enough: institutional kinds depend on some sort of public recognition. A particular piece of paper, according to this view, is money in virtue of being conferred or assigned the status of money, or of being recognized as money in a given society at a given time (e.g. Lagerspetz, 1989; Searle, 1995; Celano, 1999; Thomasson, 2003).

An institutional status may be assigned directly or indirectly. As an example of the first kind, consider a situation in which an individual is unanimously considered the captain of the football team. The status may be directly conferred by forming the relevant beliefs, and perhaps by stating them verbally in a public context ("From now on, X will be the captain of our team"). Or it may be mediated by a rule that identifies some property or condition that X must satisfy in order to have that status. For example, it may be generally accepted that the most experienced player is the captain of the team. X then would have that status derivatively, in virtue of the fact that she has the relevant property (Thomasson, 2003; Khalidi, 2015).

5.1.2 Money in Folk Classification

These philosophical theories, taken at face value, have testable implications concerning the psychology of money. It might be the case, for example that folk categorization processes reflect mainly functional considerations—that ordinary people consider money whatever fulfills the function of medium of exchange, store of value, and unit of accounting in a given society. Alternatively, the folk might classify something as money only if it has been assigned the status of money. Or it might be that people's taxonomic judgments are driven by the possession of properties that are considered necessary for the fulfilment of the functions of money. Whether any of these accounts is psychologically valid, of course, would be an empirical matter.

Understanding folk classification is interesting for various reasons. If we start from the assumption that science offers a truthful description of the world, then empirically studying the processes that govern folk classification may be helpful in preventing certain systematic errors that people make. This possibility has been taken seriously in the case of folk biology, with research considering how folk classification might hamper philosophical and scientific scholarship (Leslie, 2013) as well as education (Gelman & Rhodes, 2012). The social sciences, however, do not enjoy the same degree of respect as physics or biology. One may therefore argue—as some philosophers do—that current economic theories do not provide a reliable benchmark for the identification of errors concerning money (e.g. Searle, 1995).

But exactly the same skepticism may be directed against philosophical theories: how do we know that conferralist accounts, for example, provide an accurate picture of the nature of institutional objects? Philosophers have traditionally made heavy use of intuitions to resolve controversies in metaphysics. Although the role of intuitions has received a great deal of scrutiny over the past two decades (e.g. Alexander and Weinberg, 2007; Knobe & Nichols, 2007), in the case of institutional kinds this method can be justified in a principled manner: if the existence of institutional kinds depends on people's classificatory practices, then what the folk take to be the case would indeed be the case. Intuitive judgements, in other words, could provide genuine insights on the nature of institutional entities because our judgments would be *constitutive* of institutional reality (Ruben 1989; Thomassson 2003). Alternatively, the coordinative function of monetary objects might be fulfilled independently of people's awareness: folk theories may in principle diverge significantly from the psychological mechanism that sustain and reproduce institutions (Guala, 2016).

We do not intend to solve these thorny meta-theoretical issues here. We would simply like to point out that, regardless of one's philosophical inclinations, the empirical study of folk classification may deliver plenty of information that (directly or indirectly) bears on debates concerning the ontology of money.

5.1.3 Recent Work on the Psychology of Money

Unfortunately, we know relatively little about how people represent institutional kinds. Studies of artifacts have revealed that their physical properties, affordances, and intended functions play a role in categorization processes (see e.g. Chaigneau et al., 2004, and Margolis & Laurence, 2007, for overviews). And there is some evidence that people may classify institutional objects in a way that is broadly analogous to how they classify ordinary artifacts (see Roversi et al., 2013; Villani et al., 2022; and, for theoretical analysis, Tummolini & Castelfranchi, 2006). However, we also know that people believe that an individual can use dollar bills to purchase goods because of the current social position of dollar bills, rather than their physical structure (Noyes et al., 2018). The theory (or theories) underlying the concept of money may have primarily a coordinative function, rather than a predictive/explanatory function as the concepts we use to describe physical or biological objects. Therefore, people's beliefs about how money works may differ significantly from their beliefs about standard artifacts like hammers.

An important insight that has emerged from contemporary research is that people appear to represent certain categories as reflecting non-obvious causal processes (Bloom, 1996; Keil, 1989; Gelman, 2003, 2013; Malt & Sloman, 2007). Natural kinds provide the best evidence for these intuitive causal theories; in particular, people represent animal categories as possessing internal essences that cause their properties (Gelman, 2003; Medin & Ortony, 1989). For example, after observing an individual squirrel building a nest, one may infer that other squirrels build nests too, based on the assumption that squirrels share a fundamentally similar biology that causes their behavior. From this perspective, categories like *money* or *banker* may appear less intuitively helpful for making sense of the world, as they lack a unique internal or inherent structure. Dollar bills are mere pieces of paper, while a credit card may be a piece of plastic or just a serial number linked to a bank account. Similarly, there is nothing 'deep' about an individual that makes them a banker.

Therefore, one may conclude that categories like banker and money are not guided by intuitive theories (Keil, 1989).

However, categories can also be constituted by *external* structures, such as norms, conventions, or other regularities in the social world. Although social structures cannot be directly observed, they can sustain correlated clusters of properties. For example, people may infer that dollar bills can be used to purchase goods from the position they occupy in society rather than from their physical properties (Noyes et al., 2018). When one observes a dollar bill, one cannot see any coordinated social processes; nevertheless, one may use information about physical and social structures to infer various properties of dollar bills, such as their exchange or their store value.

Indeed, we know that people reason about many social categories and category-linked properties in virtue of external social processes: they believe for example that individuals constitute groups when they mutually intend to be a group (Noyes & Dunham, 2017, 2020; Noyes & Keil, 2020). Therefore, an individual is perceived to belong to a group not because she is similar to other members of the group but because she occupies a particular position in the group's stable pattern of social relationships (Rhodes, 2013). When hearing generic statements like "boys play with trucks," people can reason that the category-property relationship obtains because of social-structural processes like cultural constraints on the preferences boys can express, rather than because there is anything inherent or intrinsic about boys (Noyes & Keil, 2019, 2020; Vasilyeva et al., 2018; Vasilyeva & Lombrozo, 2020). Therefore, people may possess intuitive theories of social kinds as well, but ones that locate the causal system externally rather than within individual members.

5.1.4 Empirical Approach and Rationale

Our first hypothesis was that people categorize objects as money based on their current position within society regardless of their material features or their original intended purpose. For example, according to this hypothesis, participants should categorize cigarettes or shells as money if they came to occupy that position in society, even though shells or cigarettes were not designed for that purpose. This type of category judgment uniquely relies on stable features of the external social structure rather than the structure of the object *per se.* This would accord with previous evidence that participants believe the function of money is determined by its position within external systems of rules, rather than its physical properties (Noyes et al., 2018). This causal relationship differs from how the functions of standard artifacts (like chairs and hammers) work, which depend on physical structure rather than social structure.

However, we also conjectured that the original intended purpose of an object could contribute to money category judgments. Although historical, defunct currency objects like Roman coins no longer Occupy the position of money in society, their material features cannot be explained without acknowledging their original intended purpose. In this regard, people may persist in categorizing these objects as money. This manner of categorization would be noteworthy because money does not even need to involve currency objects. For example, many communities are becoming increasingly "cashless," such that money is realized entirely by digital records and not by material objects. Nevertheless, insofar as currency objects do exist, whether currently used or not, their existence and material properties depend on their intended purpose as money.

To test these hypotheses, we devised vignettes that doubly dissociated the current social position of an object from its intended purpose. Objects that were originally created to be money can stop occupying that position in society for several reasons: because of the collapse of the society that sustained them (e.g., Roman coins), because of regime change (e.g., the shift to Confederate States dollars during the Civil War), or because of ordinary currency turnover (e.g., discontinued currency objects in the United States). In Study 1, we examined a hypothetical collapse of American society. This allowed us to test people's concepts of dollar bills as a defunct currency object. In this scenario, people begin to use shells and cigarettes as mediums of exchange. These objects allow us to isolate the current social position from their intended purpose: shells lack an original purpose (as natural objects) and cigarettes had an alternative original purpose. In Study 2, we examined a regime change involving a novel, fictional society so that we could isolate the intended purpose of a currency object from people's familiarity with its use as money. We expected people to categorize the cigarettes and shells as money when they occupied that position in the community. Of interest was the extent to which people might also categorize defunct currency objects as money.

In Study 3, we asked: what does it mean for an object to occupy the position of money in society? The philosophical literature provides at least two types of perspectives on this question: (1) A community's coordinated behavior with respect to an object (or verifiable object) realizes the social functions of money: being a medium of exchange, store of value, and a unit of account (Guala, 2016). Thus, an object is money if it objectively realizes these social functions regardless of whether people recognize it as money. (2) A community collectively recognizes an object (or verifiable record) as having the social status of money (Searle, 1995, 2010). This status is associated with deontic properties, such as privileges and obligations: e.g., by being recognized as money, an individual is allowed to use the objects to settle debts. Thus, an object is money if people recognize it as having this status.

As a first approximation, both of these accounts agree that money is a social kind—money is not determined by physical structure but by social structure. And these factors may be causally related in complex ways: for example, the fact that a currency has been recognized as legal tender may automatically lead people to infer that it is used for trading purposes. And the fact that a currency has been historically created to be money, may be used to infer that it is still used and recognized as such in a given society. Study 3 was designed to investigate the extent to which people rely on, and prioritize, collective recognition, functional coordinated behavior, or history (intended function) in their category judgments. In principle, these factors are all dissociable: one-cent Euro coins, for example, are recognized as money but rarely function as mediums of exchange in practice. The opposite combination occurs too: because cash is often banned in correctional facilities, cigarettes, postage stamps, and ramen packages have served as ad hoc mediums of exchange. It is possible that members of jail communities recognized those objects as money; but those objects could function as mediums of exchange whether they were recognized as money or not. Moreover, some currencies (like McKinley bills in the USA) have been produced in the past to fulfil money-like functions but have rapidly fallen out of circulation, while others—whether or not they were ever recognized as such—are used in informal transactions or black markets.

To explore how folk classification is affected by these factors, we used objects like one-cent Euro coins, or cigarettes in prison communities, to determine if participants rely more on collective recognition, history, or functional coordinated behavior. Past evidence suggests that people believe that both collective recognition and coordinated behavior make it possible for an individual to use an object to make a purchase (Noyes et al., 2018). One study found that participants relied more on collective recognition than function to determine if individuals constituted a group (Noyes & Keil, 2020); but even here, function partly contributed to judgments about groups. Overall, then, we expected participants to prioritize collective recognition, but we expected that function and history would still contribute to money category judgments.

Finally, we were interested in whether collective recognition and intended purpose act as causally central properties (Ahn, 1998; Ahn et al., 2000). As we have mentioned, the original intended purpose of an object provides an explanation for why it exists and why it has its material properties. Likewise, the current social position of an object explains certain other properties: for example, a dollar bill is valuable in virtue of its recognized status within society. So, we were interested in the extent to which we could identify an explanatory role of both collective recognition and intended purpose in how people make sense of properties associated with money.

To test this, we had participants generate common properties of money. We then assessed whether these properties were explained by collective recognition or intended purpose. We expected participants to endorse collective recognition as explaining the social-functional properties of money (e.g., being a medium of exchange) and intended purpose as explaining the material properties of money (e.g., depicting Presidents). To make sense of why collective recognition is more important than intended purpose, we also examined the extent to which the properties they accounted for were considered explanatory or contingent (Prasada & Dillingham, 2006, 2009). Properties vary in the extent to which people represent them as properties of the kind *per se* or properties that are contingently associated with the kind. We expected participants to rate the social-functional properties of money as more central than the material properties of money. In turn, we expected collective recognition to account for the more explanatory properties of money (its social-functional properties).

5.2 Study 1

Our first experiment was designed to elicit judgments about two different cases: First, we were interested in objects that come to occupy the social position of money but that were not originally created for that role: e.g., natural objects like shells, and artifacts like cigarettes. If people categorize these objects as money, it would show that people represent money as a social position. We contrasted these items with defunct currency objects: objects that were originally created to be money but that do not currently occupy that position. There may be a pull to categorize these as money in virtue of their original intended purpose, since their existence cannot be explained through other causal processes (unlike shells and cigarettes). To test this, we created a vignette where American society collapses: before the crisis, people use dollar bills as money; after the crisis, people initially use cigarettes or shells as money, and then they switch to the other object (Fig. 5.1). This allowed us to examine how people categorize defunct dollar bills, cigarettes/shells when they occupy the position of money ('used' objects), and cigarettes/shells when they do not ('unused' objects).

5.2.1 Method

5.2.1.1 Participants and Materials

We recruited 200 experimental subjects on Amazon Mechanical Turk. Of these, 195 completed the entire survey. For Study 1, we did not collect demographic information. Studies have demonstrated that MTurk samples are more representative than undergraduate samples; they are close to representative for race and gender but skew younger, liberal, and non-religious (Berinsky et al., 2012; Burnham et al., 2018; Huff & Tingley, 2015). This study and all subsequent studies were conducted

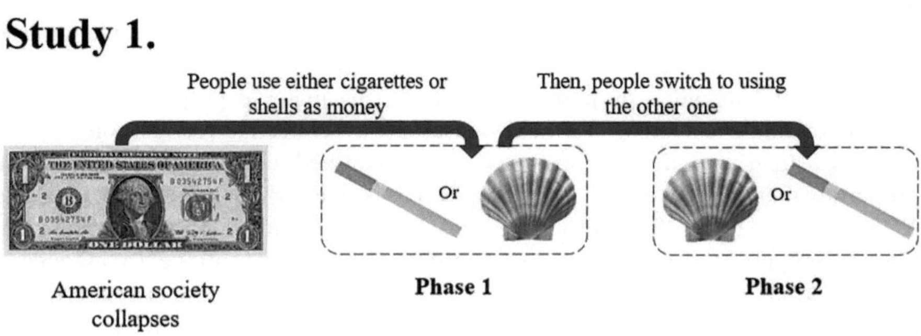

Fig. 5.1 Study 1. The vignette told participants a story about 'post-America', a society where dollar bills are no more in use, and are replaced by either cigarettes or shells as a medium of exchange

under Protocol Number 1311013027 approved by Yale's IRB Board. For all studies, we paid Connecticut minimum wage, which was $10.10 at time of testing. We paid participants 0.71$ for the 4-minute task. The full materials (including scripts, images, and structure of questions) are available on Open Science Framework: https://osf.io/6y8gf/.

5.2.1.2 Design

The task was split into two phases, corresponding to before and after a switch in which a non-currency object (cigarettes or shells) was used as a medium of exchange. In each phase, participants received three blocks of questions, presented within-subject in randomized order, corresponding to three objects: (1) *Dollar bills:* defunct dollar bills, (2) *Used:* cigarettes or shells currently acting as a medium of exchange, (3) *Unused*: cigarettes or shells not currently acting as a medium of exchange. In each block, participants expressed their agreement to three statements in randomized order: the test statement (is the object money) and two control statements.

5.2.1.3 Procedure

At the introduction of the task, participants were told they would read a story and answer questions about it. Our basic vignette involved a hypothetical scenario called 'post-America':

> American society has crumbled after a massive viral outbreak. The former government has been decimated. All of the former religious, economic, and legal systems are gone. For example, the Federal Reserve Bank and U.S. Treasury no longer exist. Descendants of that society now live among the rubble. We will call this society 'post-America.'

Subjects were told that people in post-America use peculiar objects to perform money-like functions. Depending on the experimental condition, these objects may be *cigarettes* or *shells* (this was randomized between-subject):

> People trade cigarettes [shells] for goods. People reliably expect that others will accept cigarettes [shells] for payment, so people can keep track of the cost of objects in quantities of cigarettes, and can reliably expect cigarettes [shells] to be valuable. People cannot use any other objects this way.

In each block, participants saw an image of one of the objects—a dollar bill, a cigarette, or a shell—followed by three statements that they rated on a 1–6 Likert scale ranging from "Totally disagree" to "Totally Agree":

1. *'This is money'.*
2. *'This counts as money in post-America'*
3. *'This was created for the purpose of being money'.*

We call 1 the *Basic Categorization* statement; 2 is the *Conventional Categorization* statement; and 3 is the *Purpose Categorization* statement. The judgment of interest is only the first one—what subjects consider to *be* money in post-America: dollars, cigarettes, or shells? The second and third statements, instead, are comprehension checks. Subsequently, participants read that the community switched from using shells to cigarettes or cigarettes to shells. Participants then rexamined the same three blocks of experimental statements again. The total task was very brief and could be completed in approximately four minutes.

5.2.2 Results and Discussion

Overall, we found the qualitative pattern of results was the same across phase and so we collapsed across phase for these analyses. To analyze results, we used generalized linear models with the cluster bootstrap as implemented in the R package 'ClusterBootstrap' (GLMCB; Deen & de Rooij, 2020). We report bootstrapped standard errors in parentheses. We report bias corrected and accelerated 95% bootstrapped confidence intervals and use those to draw conclusions. To investigate participants' beliefs about each of three object types (dollar bills, used cigarettes/shells, non-used cigarettes/shells), we coded the categorical variable as two contrasts: dollar bills compared to used cigarettes/shells, and used objects cigarettes/shells to unused cigarettes/shells.

First, we examined the control statements.

"Counts as" control: Participants correctly said that defunct dollar bills did not 'count as' money in the social context, $M = 1.57$ (0.08), 95%CI: [1.42: 1.75], as they rated dollar bills significantly below the midpoint of the scale (3.50). They correctly said that cigarettes and shells 'counted as' money when they were used, $M = 5.61$ (0.06), 95%CI: [5.47: 5.71], and not when they were unused, $M = 1.73$ (0.08), 95%CI: [1.58: 1.90]. Dollar bills were rated significantly lower than used cigarettes/shells by $b = -4.04$ scale points, 95% CI: [−4.28: −3.75]. Used cigarettes/shells were rated significantly higher than unused cigarettes/shells by $b = 3.88$ scale points, 95% CI [3.61: 4.11].

"Created to be" control: Participants correctly said that defunct dollar bills were originally created to be money, $M = 5.12$ (0.11), 95%CI: [4.88: 5.31]. They correctly said that cigarettes/shells were not originally created for that purpose whether they were currently used, $M = 2.10$ (0.12), 95%CI: [1.89: 2.34], or not, $M = 1.48$ (0.07), 95%CI: [1:35: 1.63]. Dollar bills were rated significantly higher than used cigarettes/shells by $b = 3.02$ (0.21) scale points, 95% CI: [2.58: 3.41], and cigarettes/shells were rated significantly higher by $b = 0.62$ (0.10) scale points when used then un-used, 95% CI: [0.44: 0.85]. Overall, participants appropriately tracked the intended purpose and current social status of the objects.

Next, we examined the critical test statement (Fig. 5.1). Participants rated defunct dollar bills significantly less as money than used cigarettes/shells: $b = -0.73$ (0.24) scale points, 95% CI: [−1.20: 0.25], and participants rated cigarettes/shells as money significantly more when they were used then unused: $b = 2.86$ (0.15) scale points, 95% CI: [2.57: 3.15]. Overall, however, participants said that dollar

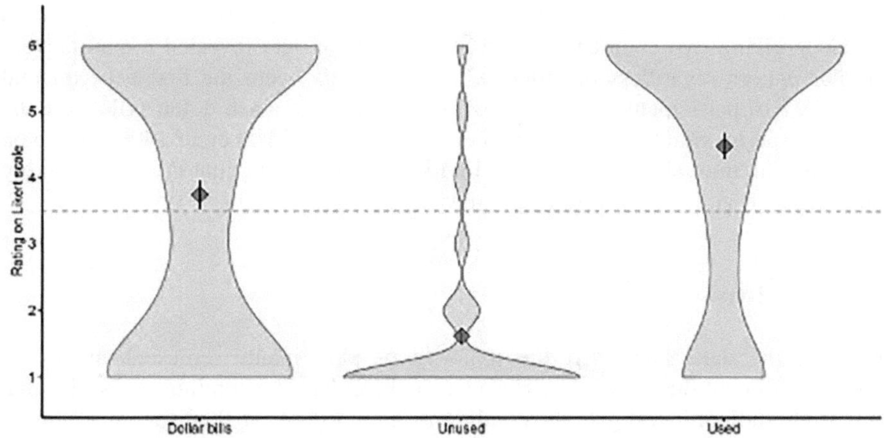

Fig. 5.2 Study 1. Participants' judgments about whether the three objects were money: Defunct dollar bills, cigarettes/shells (unused), and cigarettes/shells when acting as a medium of exchange (used). Error bars are 95% bootstrapped confidence intervals as calculated by ggplot2. Dotted line is midpoint of scale, 3.5

bills were money around the midpoint of the scale: M = 3.75 (0.15), 95%CI: [3.45: 4.03]. In fact, participants responded in a strongly bimodal fashion, as can be seen in the violin plot in Fig. 5.2, which shows the distribution of responses by the relative width. Indeed, 65.64% of participants gave a one (28.46%) or a six (37.18%) on the scale, which were the most extreme responses. This bimodality is also present, to a lesser extent, in category judgments of used cigarettes/shells; however, participants judged used cigarettes/shells as money significantly higher than the midpoint of the scale: M = 4.48 (0.13), 95%CI: [4.23: 4.72]. Indeed, participants who said that dollar bills were money were less likely to say used cigarettes/shells were money and vice versa, as revealed by a significant negative correlation: *phase one*, $r = -.49$, 95%CI: [−0.58: −0.38]; *phase two*, $r = -.53$, 95% CI: [−0.63: −0.41]. But, on closer inspection, these two response strategies were asymmetric: participants who said dollar bills were *not* money (i.e., ratings <3.50), said used cigarettes/shells were money 93.53% of the time. In contrast, participants who said dollar bills were money were bimodal in their judgments about used cigarettes/shells: 57.77% said they were money and 42.27% said they were not. Finally, unused shells/cigarettes were not rated as money, as expected: M = 1.62 (0.07), *95%* CI: [1.49: 1.78].

5.2.2.1 Order

The categorization task may imply that not every object could be money, leading to an artificial bimodal pattern. But, if that were true, then we would not expect any bimodality on the first trial, before participants were aware that they would be categorizing two contrasting cases. However, the data revealed a qualitatively similar pattern regardless of order: when dollar bills were the first experimental trial, 58% of participants gave a 1 (26%) or a 6 (33%); when dollar bills were the second experimental trial, 68% of participants gave a 1 (28%) or a 6 (40%). Even on the first trial, thus, we see substantial bimodality, suggesting that it was not merely an artifact of within-subject task demands.

5.2.2.2 Discussion

Overall, the data show that the majority of participants represent money as corresponding to the current social status an object has in a community. Notice that this study did not investigate the conditions that must be satisfied for something to acquire the status of money. Although we primarily discussed used cigarettes/shells in terms of their function as media of exchange, this study was not designed to distinguish between the collective recognition of cigarettes/shells as money and the current use of them as media of exchange—we resolve this in Study 3. In either case, participants are representing money as embedded in a community's social structure—their collective intentions and behaviors. This representation of money differs from classic essentialist representations, which would locate money-hood within the intrinsic or internal features of an object rather than in the social features of the surrounding context. But this representation of money also differs substantially from many accounts of standard artifacts, where participants tend to prioritize either material features of the object itself (i.e., the causal efficiency of its structure for performing a function; Chaigneau et al., 2004) or its design history (i.e. intended purpose; Bloom, 1996; Matan & Carey, 2001).

Design history, to be sure, did influence some judgments. We found that many participants continued to categorize defunct dollar bills as money; moreover at least half of the participants who categorized defunct dollar bills as money did not categorize used cigarettes/shells as money. Therefore, the intended purpose of an object continued to contribute to how a significant number of participants represented money, and was even the primary or exclusive way that at least some of them represented money. Although we found that participants were tracking the control information, we could not rule out that they were tempted to categorize defunct dollar bills as money because of their familiarity with dollar bills as money, rather than the intended purpose of these objects. Therefore, we conducted a follow-up study using a novel currency object. If participants categorize defunct dollar bills as money because of their prolonged exposure to dollar bills as money, then we should see this response strategy disappear when we use entirely novel currency objects. But, if participants represent defunct dollar bills as money in virtue of their original intended purpose, then we should see this response strategy persist.

5.3 Study 2

We have seen that people continued to categorize defunct dollar bills as money: is this merely because participants are familiar with dollar bills as money (through prolonged exposure, and their current status as money in U.S. society), or are people basing their category judgments on original intended purpose? We resolved this question by investigating a replication of Study 1 using a novel currency object. This allowed us to isolate the contribution of intended purpose to category judgments.

In designing this study we also wanted to check whether the responses were correlated with some individual characteristics of the respondents. For example, the level of education –economics education, in particular—might be correlated with the category judgments elicited in the experiment. A student may have heard during a lecture or read in a book that money is whatever fulfils the function of medium of exchange, store of value, and unit of accounting. It is plausible that such an individual would identify used objects as money in the post-America scenario, while a subject who has not been exposed to economics education would not.

The broad cultural environment in which individual respondents have acquired their concepts of money may provide other explanatory factors. There is evidence, for example, that children who have received a religious education are more likely to have 'essentialist' beliefs about social properties such as ethnicity, race, and gender (e.g. Diesendruck & Haber, 2009). Other studies report a correlation between socio-economic status (which often reflects differences in formal education) and the conceptualization of social class (e.g. Kraus & Keltner, 2013). Finally, categorization seems to be influenced by 'ways of thinking', such as similarity-based or causality-based reasoning, that are correlated with ethnicity (e.g. Norenzayan et al., 2002). To cast the net widely, we collected data about nine demographic variables: Education, Economics education, Political orientation, Age, Gender, Race, Household income, and Childhood income.

5.3.1 Method

5.3.1.1 Participants and Materials

We apportioned 200 Hits from Amazon Mechanical Turk. 206 people completed the survey. See participant section for Study 1 for details about demographics, IRB, and pay. The full materials (including scripts, images, and structure of questions) are available on Open Science Framework: https://osf.io/6y8gf/. Aspects of the materials were adapted from Noyes et al. (2018). The novel currency object was visually depicted with a Katanga cross, which is an X-shaped iron ingot currency object used in the State of Katanga in the early twentieth century. The test questions were identical to Study 1 except the word "money" was referred to with the word "kayan."

5.3.1.2 Design

Participants were presented, within-subject, with two blocks of statements in random order corresponding to the two objects: (1) *Defunct currency object,* a defunct, novel currency object and (2) *Used shells,* natural shells currently acting as a medium of exchange. In each block, participants examined three statements in randomized order: the test statement (Is the object 'kayan') and two control statements.

5.3.1.3 Procedure

At the introduction of the task, participants were told they would read a story and answer questions about it. Participants then read an introduction to a novel island community (called "Vawnsie") and learned about "kayan," which was the novel island community's money:

> On an island far away there exists a group of people called Vawnsies. Vawnsie society is unlike our society in almost all ways. They speak a different language, have different family structures, and have a variety of cultural and religious practices very unlike our own. On Vawnsie Island there is something called 'Kayan'. On Vawnsie Island, people trade the objects depicted below for goods. People expect that others will accept them for payment. This allows people to keep track of the cost of goods in quantities of these objects. People can expect them to remain valuable. People cannot use any other objects this way.

This description introduced them to novel currency objects, which were visually depicted. Participants then read a vignette describing a regime change on the island, during which the community stopped using the currency object to trade for goods and services and switched to using natural shells (Fig. 5.3). Participants then examined two blocks of experimental statements. In each block, participants saw an image of one of the objects followed by three statements that they rated on a 1–6 Likert scale ranging from "Totally disagree" to "Totally Agree." The test statement was: "This is kayan." The control statements were: "This counts as kayan on Vawnsie Island now," and "This was created for the purpose of being kayan."

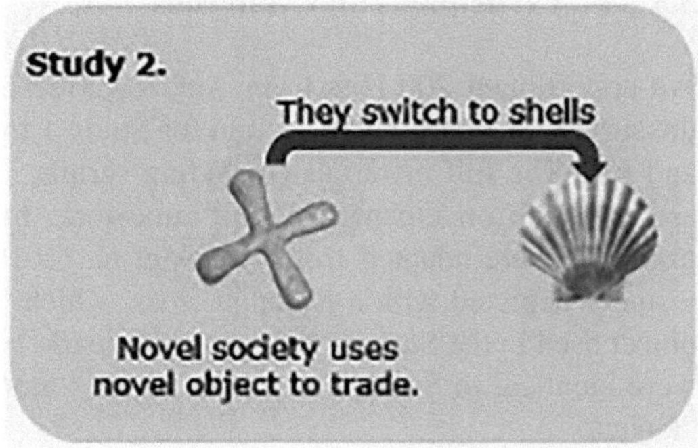

Fig. 5.3 Study 2: the vignette told participants a story about "Vawnsie Island", a society where X-shaped objects were used as "kayan" (money) but are later replaced by shells

5.3.2 Results and Discussion

Participants reacted correctly to the control statements: they said that the defunct currency objects did not ‘count as’ money in the social context, $M = 2.71$ (.14), 95%CI: [2.44: 3.00], and they correctly said that used shells did, $M = 4.97$, 95%CI: [4.74: 5.16].[1] The defunct currency objects were rated significantly lower by $b = -2.26$ (.21) scale points, 95%CI: [−2.63: −1.81]. Participants also correctly said that the defunct currency objects were originally created to be money, $M = 4.92$ (.09), 95%CI: [4.73: 5.09], and correctly said that the used shells were not, $M = 2.49$ (0.12), 95%CI: [2.25: 2.75]. The defunct currency objects were rated significantly higher by $b = 2.43$ (0.17) scale points, 95% CI: [2.07: 2.73]. Taken together, participants appropriately tracked the intended purpose and current social status of the objects.

Next, we looked at the critical test statement (Fig. 5.2). Participants rated the novel, defunct currency object as money significantly lower than used shells by $b = -0.64$ (0.24) scale points, 95%CI: [−1.10: −0.15]. As before, participants were overall around the midpoint of the scale in their judgments about defunct dollar bills: $M = 3.68$ (0.14), 95%CI: [3.39: 3.95], which as before reflected an underlying bimodal response strategy (see Fig. 5.4). Also, as before, participants

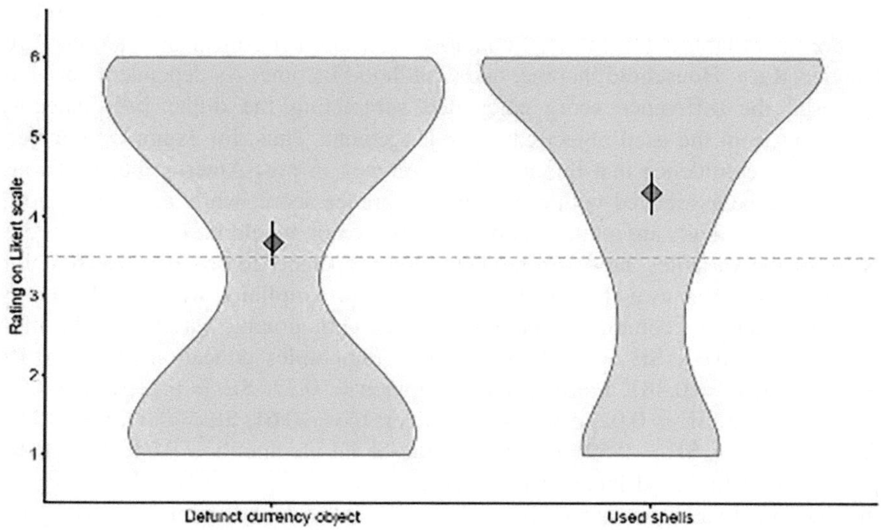

Fig. 5.4 Study 2. Participants judgments’ about whether the two objects were money: A novel, defunct currency object and shells currently acting as a medium of exchange. Error bars are 95% bootstrapped confidence intervals as calculated by ggplot2. Dotted line is midpoint of scale, 3.5

[1] As in Study 1, we used generalized linear models with the cluster bootstrap as implemented in the R package ‘ClusterBootstrap’ (GLMCB; Deen & de Rooij, 2020). We report bootstrapped standard errors in parentheses. We also report bias corrected and accelerated 95% bootstrapped confidence intervals and use those to draw conclusions about statistical significance.

did consistently judge that used shells were money, $M = 4.32$ (0.13), 95% CI: [4.05: 4.56], reflecting an attenuated bimodal response pattern such that the majority of participants were saying the shells were money. As in Study 1, there was a significant negative correlation between responses to defunct currency objects and used shells, $r = -0.51$ (.053), 95%CI: [−0.62: −0.41]. Participants who said that defunct currency objects were *not* money, said that used shells were money 92.22% of the time. Participants who said that defunct currency objects were money, were bimodal in judgments about used shells, saying that used shells were money 55.56% of the time and not money 44.44% of the time.

The currency study thus closely replicated Study 1 despite using a completely novel currency object and a regime change rather than a societal collapse. Once again, the majority of participants represented money as embedded in societal structure. But we also replicated the finding that many participants categorize defunct currency objects as money, suggesting that intended purpose plays an additional or exclusive way of representing money for some participants. The persistence of this effect in a novel context shows that participants were not categorizing the defunct currency objects as money because of their familiarity with the object as money. Instead, many participants believe that intended purpose is sufficient to make an object money. Across both studies, roughly 40% of participants exclusively relied on current social status, 30% relied on current social status and intended purpose, and 25% relied exclusively on intended purpose.

Finally, we looked for correlations between category judgments and nine socio-economic variables: Education, Economics education, Political orientation, Age, Gender, Race, Household income, and Childhood income. As dependent variable, we used the difference score calculated subtracting the dollar bills category judgment from the used object category judgement. Thus, for example, a subject expressing confidence that dollar bills are money in post-America but shells are not, would be associated with a negative difference score; while a subject who is confident that shells are money but dollar bills are not, would have a positive score. None of the variables, however, predicted the propensity to use effective function or conferred status as a classification criterion. For simplicity, we report here only the statistics of the combined model with all the demographic variables: Education level: $b = -0.03$, $SE = 0.22$, $p = 0.903$. Economics education: $b = -0.19$, $SE = 0.27$, $p = 0.481$. Political Orientation: $b = 0.37$, $SE = 0.21$, $p = 0.084$. Age: $b = 0.02$, $SE = 0.02$, $p = 0.405$. Gender: $b = -0.61$, $SE = 0.49$, $p = 0.214$. Race: $b = 0.10$, $SE = 0.59$, $p = 0.865$. Household income: $b = 0.04$, $SE = 0.18$, $p = 0.840$. Childhood income: $b = 0.18$, $SE = 0.17$, $p = 0.315$. This suggests that judgment variations cannot be explained by individual differences—not by the plausible properties that we have checked in our study, at least.

5.4 Study 3

In Study 1, objects like cigarettes or shells may be perceived as occupying the position of money because they are collectively recognized as money, but also because they fulfil the effective function of money. Are both of these conditions sufficient for participants to categorize an object as money or do they play different roles in people's concepts of money? To answer this question, we can examine (1) objects that are recognized as money but are not used in practice, and so do not effectively function as mediums of exchange; (2) objects that act as mediums of exchange in society even though they are not recognized as money by the community. In Study 3, we investigated these two cases in a full factorial design that varied the recognition of an object as money, its effective function as a medium of exchange, and (as in the previous studies) its original intended purpose.

5.4.1 Method

5.4.1.1 Participants and Materials

We apportioned 200 Hits from Amazon Mechanical Turk. 198 people completed the survey. See participant section for Study 1 for details about demographics, IRB, and pay. The full materials (including scripts, images, and structure of questions) are available on Open Science Framework: https://osf.io/6y8gf/.

5.4.1.2 Design

Participants judged whether 8 objects were money: these 8 objects represented every cell in a 2 by 2 by 2 factorial design combining *History* (Originally made to be money or not), *Function* (Currently acting as a medium of exchange or not), and *Recognition* (Collectively recognized to be money or not). All participants received every object in randomized order.

5.4.1.3 Procedure

At the introduction of the task, participants were told they would read eight descriptions of objects and would be asked a single category judgment. Participants received eight blocks in randomized order. In each block, participants first read a brief description of the object they were being asked to consider. The basic template of the description was the same in each block:

- *Historical information*: X were [were not] made for the original purpose of being a mode of payment. They originally counted [did not count] as a mode of payment in American society.
- *Recognition information*: Local norms in America stipulate [do not stipulate] that X are a mode of payment. The government stipulates [does not stipulate] that X are a mode of payment.
- *Function information*: In practice, X are used [are not used] as a mode of payment. People trade [do not trade] X for goods and services.

For example, in one block subjects read a story about a currency that had once been used as money, but is not currently recognized and is not currently used as such:

> 500 dollar bills depicting William McKinley were used in the 1930s but became obsolete. Now the bills are only housed in museums. McKinley bills were made for the original purpose of being a mode of payment. McKinley bills originally counted as a mode of payment in 1930s America. Local norms in America do not stipulate that the McKinley bills are a mode of payment. The government does not stipulate that the McKinley bills are a mode of payment. In practice, McKinley bills are not used as a mode of payment in America. People never trade McKinley bills for goods and services.

Participants then read three tightly controlled statements that reinforced whether the currency was originally created to be money or not, was currently acting as a medium of exchange or not, and was currently recognized as a medium of exchange. These statements were presented in randomized block. Participants were then asked: "Are these objects money," which they rated on a 1–6 Likert scale ranging from "Definitely not" to Definitely yes."

5.4.2 Results and Discussion

First, we analyzed the results using a multi-level model with History, Function, and Recognition as predictors (Fig. 5.5).[2] All three of these factors significantly contributed to money category judgments: there was a significant effect of History, $b = 1.14$ (.08), 95%CI: [0.99: 1.30], a significant effect of Function, $b = 1.22$ (.07), 95%CI: [1.09: 1.36], and a significant effect of Recognition, $b = 2.03$ (.09), 95%CI: [1.85: 2.19]. Therefore, an object was more likely to be categorized as money if it was originally made to be money, was currently used as a medium of exchange, or was collectively recognized as money, relative to an object that had none of these features (e.g., a natural shell on a beach). History and Function affected money category judgments to a similar extent, whereas Recognition had a significantly

[2] We analyzed the results as in Study 1: we used generalized linear models with the cluster bootstrap as implemented in the R package 'ClusterBootstrap' (GLMCB; Deen & de Rooij, 2020). We report bootstrapped standard errors in parentheses. We also report bias corrected and accelerated 95% bootstrapped confidence intervals and will use those to draw conclusions about statistical significance.

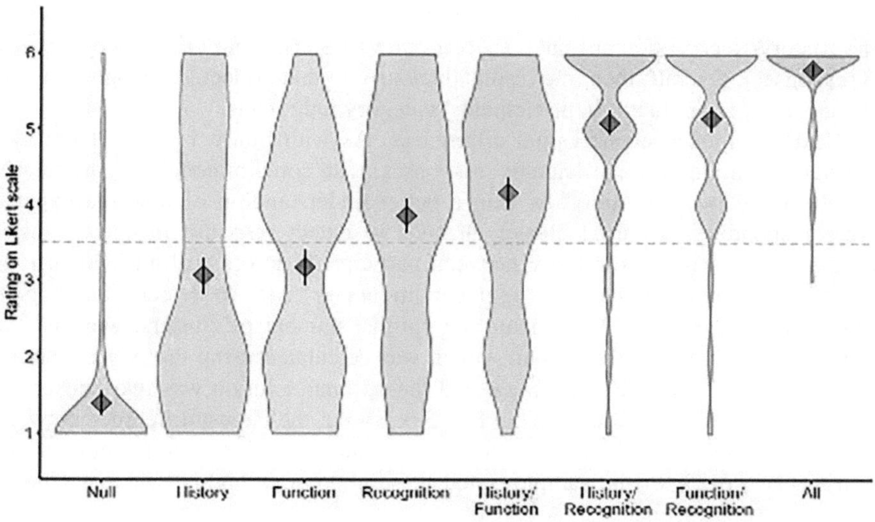

Fig. 5.5 Study 3. Participants judgments' about whether an object was money by whether it was originally created to be money (History), currently used as a medium of exchange (Function), or collectively recognized as money (Recognition). Error bars are 95% bootstrapped confidence intervals as calculated by ggplot2. Dotted line is midpoint of scale, 3.5

greater effect (as revealed by the non-overlap of the confidence intervals). Next, we added the interaction terms to the model. The three-way interaction effect was non-significant: $b = .10$ (.20), 95%CI: [−0.30: 0.48]. But each of the two-way interaction terms was significant: History*Function, $b = -0.68$ (0.14), 95%CI: [−0.95: −0.39]; History*Recognition, $b = -0.44$ (.17), 95%CI: [−0.77: −0.12], Function*Recognition, $b = .48$ (.15), 95%CI: [−0.77: −0.20]. Each of these two-way interactions demonstrate (unsurprisingly) that the difference between objects with a single property and the control object (which had none) was larger than the difference between objects with one property and objects with two properties.

Next, we compared the conditions to the midpoint. First, we looked at the singleton conditions where the object had only one of the three properties. Merely being recognized as money was sufficient to be categorized as money: $M = 3.86$ (0.11), 95%CI: [3.64: 4.07]. But, merely being used as a medium of exchange was not, $M = 3.17$ (0.11), 95%CI: [2.96: 3.39]. Nor was being originally created to be money sufficient on its own, $M = 3.07$ (0.13), 95%CI: [2.82: 3.32]. Objects that had two properties were all rated above the midpoint: History-Function: $M = 4.17$ (0.10), 95%CI: [3.96: 4.36]; History-Recognition: $M = 5.09$ (.08), 95%CI: [4.91: 5.25]; Function-Recognition: $M = 5.15$ (.08), 95%CI: [4.98: 5.30]. Therefore, none of the three properties was necessary to be categorized as money. Notably, however, objects recognized as money (History-Recognition, Function-Recognition) were rated higher than objects that were not recognized (History-Function). From the violin plots we can see that a bulk of participants said the History-Function object was partially money (most responses at 4 and 5), whereas the majority participants said the History-Recognition or Function-Recognition objects were fully money (mostly 6 responses). Overall, then, the results demonstrate that collective recognition was the most important factor in participants' category judgments.

Next, we looked at individual differences. As with Study 1 and 2, there was substantial variation in participants' responses, with some bimodality (particularly for the Function condition). To gain a better understanding of how participants were responding, we used cluster analysis to detect response profiles. Cluster analysis measures the similarity between participants across multiple dimensions (in this case, responses to the eight conditions) in order to detect clustering in that dimensional space. To estimate the optimal number of clusters, we used the optimum average silhouette width, which was calculated using the *pamk* function from R's "fpc" package. This suggested the optimal solution was two clusters of participants, which we visualized in Fig. 5.6. By far, the largest difference between these two samples was their responding to the Function and History-Function trials: Cluster Two (42% of the sample) rated these conditions highly, whereas Cluster

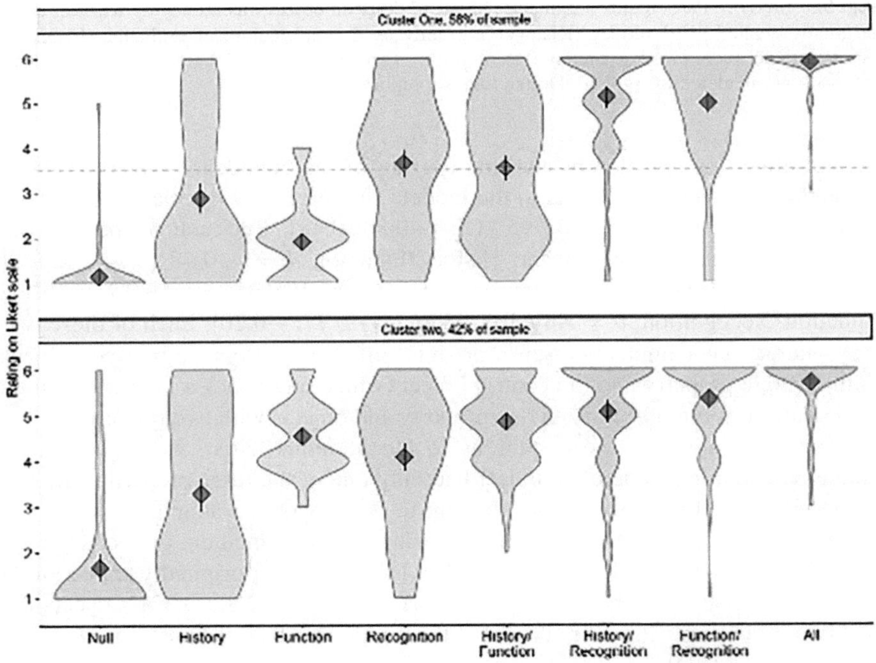

Fig. 5.6 Study 3 divided by two response clusters. Participants judgments' about whether an object was money by whether it was originally created to be money (History), currently used as a medium of exchange (Function), or collectively recognized as money (Recognition). Error bars are 95% bootstrapped confidence intervals as calculated by ggplot2. Dotted line is midpoint of scale, 3.5

One (58% of the sample) rated these conditions relatively low. In particular, Cluster Two responded that acting as a medium of exchange was sufficient to be money, whereas Cluster One responded that it was not (as visible in the violin plots). Otherwise, they were fairly similar. So, it seems that both participant clusters prioritized the current social status of an object relative to intended purpose, which is consistent with Study 1 and 2. But they differed in whether they represented current social status exclusively in terms of collective recognition or in an undifferentiated way, such that they showed little difference in their reasoning about collective recognition and function. Possibly, some participants think of functional behavior that is not recognized or formally agreed-upon as more accidental, and so as less stably predictive and explanatory. In any case, collective recognition was a far more consistent part of participants concepts of money, whereas function was only important for some participants.

As in Study 1 and 2, intended purpose (History) varied considerably in its importance to individual participants but overall contributed to money category judgments. In comparison to Study 1 and 2, original intended purpose seemed relatively less important in this setting. One major difference is that current social status in this study was divided into function and recognition, which necessarily meant that there was twice as much emphasis on the current role the object had in the community relative to its history. Likewise, this study focused on an object with the most far removed history. The defunct currency object described in the vignette in fact was discontinued in 1934. In contrast, Study 1 relied on current one-dollar bills and Study 2 described the creation of the novel currency object at the beginning of the study and presented participants with an abrupt regime change. So, Study 3 systematically deemphasized the intended purpose of objects relative to current social status, which makes it less surprising that the effect of intended purpose was slightly weaker here. Nevertheless, Study 3 adds additional evidence that original intended purpose influences money category judgments to a nontrivial extent.

One possible explanation of this effect is that institutional objects are classified according to their *disposition* to fulfil a specific function, such as medium of exchange, in a given society. An object that has been used as money in the recent past, then, might be perceived as more "money-like" than an object that has not fulfilled that function for a long time—or not at all—in that specific society. Since our data do not allow to test the dispositional hypothesis, however, we offer this remark as a suggestion for future research.[3]

---

[3] We are grateful to a reviewer for suggesting this possible explanation.

5.5 Study 4

In sect. 1 we have suggested that participants categorize money through causal-explanatory reasoning. For example, participants may categorize defunct currency objects as money because their original intended purpose explains their properties. This hypothesis, however, has not been directly tested so far: does causal-explanatory reasoning actually play a role in people's category judgments? To see whether this account is plausible we investigated the properties people generate for money and how they explain those properties. A first group of participants generated three properties of money. Two additional groups of participants evaluated the twenty most frequently generated properties: they evaluated whether the properties were explained by the community's current rules/practices and the creators' original intentions. They also evaluated whether the properties were central or not ("principled", in the terminology of Prasada & Dillingham, 2006). Core properties are treated as properties of the kind *per se* (e.g., "dogs have four legs"), whereas associated properties are treated as incidentally generalizing to the kind (e.g., "dogs wear collars"). This study applied the distinction to a set of properties that were specifically generated for the concept of money.

5.5.1 Method

5.5.1.1 Participants and Materials

For the property generation task, we apportioned 200 Hits from Amazon Mechanical Turk (195 people completed the survey). For the subsequent property evaluation tasks, we apportioned 200 Hits from Amazon Mechanical Turk for the forced choice version (195 completed the survey) and 200 Hits for the continuous version (188 completed the survey). See participant section for Study 1 for details about demographics, IRB, and pay. The full materials (including scripts, images, and structure of questions) are available on Open Science Framework: https://osf.io/6y8gf/.

5.5.1.2 Design

A first group of participants generated properties about money (and completed a replication of Study 1). Two groups of participants subsequently evaluated the 20 most frequent of those properties on either forced-choice or continuous measures. *Property generation + replication*: participants first generated three properties of money and then they did a shortened form of Study 1. Participants received three blocks of questions, presented within subject in randomized order, corresponding to three objects: (1) *Dollar bills:* defunct dollar bills, (2) *Used:* cigarettes or shells currently acting as a medium of exchange, (3) *Unused*: cigarettes or shells not currently acting as a medium of exchange. *Property Evaluation:* participants received two blocks presented within subject in randomized order: a block where they selected (forced choice) or rated (continuous) explanations of the 20 properties, and a block where they selected (forced choice) or rated (continuous) whether the property was central or not.

5.5.1.3 Procedure

Property generation + replication: the procedure was the same as Study 1, except (1) after reading the study introduction but before reading the vignette about the collapse of American society, participants were asked to enter three properties of money into three textboxes. (2) There was no second phase, and so participants did not see a switch from shells to cigarettes or cigarettes to shells. *Property Evaluation:* participants were told they would evaluate 20 properties of money. They then received the explanation block and core-property block. In the explanation block, participants saw 20 trials where a picture of a dollar bill was presented with a statement about a single property: e.g., "This is a medium of exchange." In the forced choice version, they were asked "Why" with three options "Because of the community's current rules and practices," "Because its creators decided to make it that way," "Because of something else." In the continuous version, they evaluated the first two explanations on a 1–6 Likert scale ranging from "Completely disagree" to "Completely agree." In the core property block, participants again saw a picture of a dollar bill with a statement about a single property. They were then asked to evaluate whether the property was a central one. They read (for example): "This is durable because it's money: money, by virtue of being the kind of thing it is, is durable." In the forced choice version, they selected true or false. In the continuous version, they rated the statement on a 1–6 Likert scale ranging from "Completely disagree" to "Completely agree."

5.5.1.4 Coding of Properties

For many properties, the entries were exactly the same. For example, many participants wrote in the single word "green." But, for other cases we pooled together entries that were highly related: for example, "you buy things with it," "medium of exchange," "tradeable" were counted as a single property "medium of exchange." We decided to move the 20 most frequently generated properties into the property evaluation task: this set contained every property generated at least five times and encompassed 80% of the total entries.

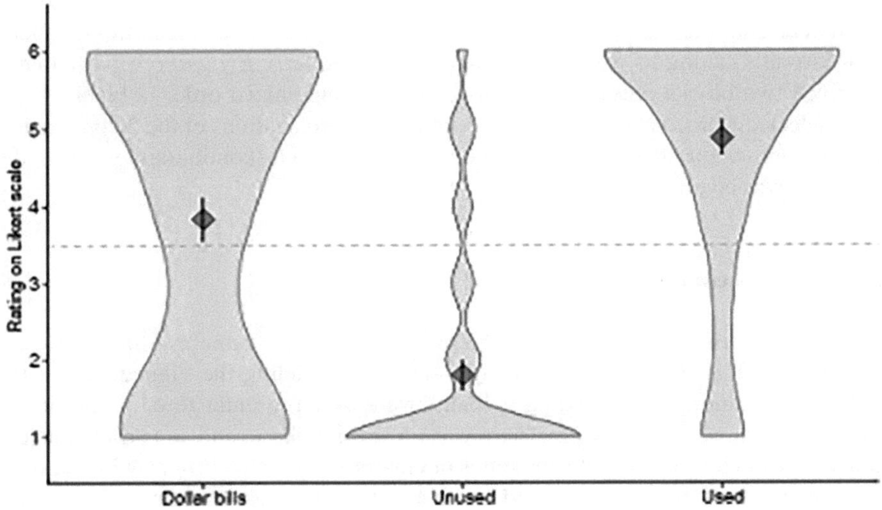

Fig. 5.7 Study 4: replication of Study 1. Participants' judgments about whether the three objects were money: defunct dollar bills, cigarettes/shells (unused), and cigarettes/shells when acting as a medium of exchange (used). Error bars are 95% bootstrapped confidence intervals as calculated by ggplot2. Dotted line is midpoint of scale, 3.5

5.5.2 Results and Discussion

5.5.2.1 Replication of Study 1

Overall, we found a qualitatively similar pattern of results as Study 1 (Fig. 5.7). Participants rated defunct dollar bills, $M = 3.84$ (1.44), 95%CI: [3.54:4.11], and used cigarettes/shells, $M = 4.88$ (.11), 95%CI: [4.62: 5.08], as money significantly higher than the midpoint of the scale. They rated defunct dollar bills as money significantly lower than used cigarettes/shells by $b = -1.04$ (.22) scale points, 95%CI: [−1.47: −0.60]. We found a similar bimodal response pattern on category judgments of defunct dollar bill, which was significantly negatively related to judgments of used cigarettes/shells, $r = -.44$ (.05), 95%CI: [−0.53: −0.33]. Participants who said defunct dollar bills were *not* money said used cigarettes/shells were 98.7% of the time. Participants who said defunct dollar bills were money said cigarettes/shells were 73.50%. Therefore, roughly 45% of participants said both were money, roughly 40% said only used cigarettes/shells were, and only 16% said only defunct dollar bills were. Overall, then, we replicated the results of Study 1—particularly the importance of current social status to money –but fewer participants relied exclusively on intended purpose and more participants relied on both intended purpose and current social status.

5.5.2.2 Properties

Next, we examined the relationship between explanations and centrality by property. For these models, the property is the unit of analysis (N = 20), with values reflecting the average rating from participants. Because explanation and core-property ratings were not within-subject, we relied on GLM with normal, non-clustered bootstrap. First, we looked at the forced choice version of the task. In a model with current rules/practices as predictor and centrality as outcome, we found a significant positive relationship: $b = 0.43$ (.07), 95%CI: [0.25: 0.54]. In a model with the creator's original intentions as predictor and centrality as outcome, we found no significant relationship, $b = -0.28$ (0.12), 95%CI: [−0.47: 0.04]. Second, we looked at the continuous version of the task. In a model with current rules/practices as predictor and core-property (centrality) as outcome, we found a significant positive relationship: $b = 0.68$ (0.12), 95%CI: [0.41: 0.87]. In a model with creator's original intentions as predictor and centrality as outcome, we found no significant relationship: $b = 0.27$ (33), 95%CI: [−0.54: 0.68]. Therefore, *the community's rules/practices explain more core properties, whereas the creator's original intentions do not.*

Using the forced choice and continuous rating of the two explanations, we looked at how the twenty properties clustered. The properties belonged to three clusters: One cluster was the social-functional properties: Serving as a medium of exchange, being valuable, being divisible, being useful, being a unit of account, being necessary, being in limited supply, and having stable value. These properties were best explained by the community's current rules/practices on the forced choice ($M = 67.88\%$ vs. 25.37%) but were rated high on both of the continuous measures ($M = 4.78$ vs. 4.63). A second cluster was material properties: being made of paper, being green, being easy to carry, being durable, having ink, being tangible, depicting Presidents, being soft/flexible, having uniform appearance, being rectangular, and having a serial number. These properties were best explained by the creator's original intentions on the forced choice ($M = 76.36$ vs. 17.18%) and the continuous measure ($M = 5.07$ vs. 3.78). A final cluster was the lone property "dirty." We found that social-functional properties were rated as significantly more central than material properties, $b = 0.21$ (0.04), 95%CI: [0.12: 0.28] (Table 5.1).

These results carry interesting information on how participants represent money. Foremost, these data confirm our premise that money has a *dual structure* combining the functional role it serves within a community and the material features of the currency objects that instantiate that role. Participants rated social-functional properties as more characteristic of money than material properties, which helps to show that *money as a kind is primarily understood as a social-functional kind* rather than a kind of physical object. These social-functional properties were best explained by the community's *current rules and practices*; indeed, the more participants said a property was 'core' the more they said it was explained by the community's current rules and practices. But, as these data also make clear, *material properties* are often brought to mind when people think about money.

Table 5.1 Frequency and ratings of the 20 most frequently generated properties

		Forced Choice				Continuous		
Property	Count	*Rules/practices*	*Creator*	*Other*	*Centrality*	*Rules/practices*	*Creator*	*Centrality*
Exchange	75	0.69	0.28	0.03	0.96	5.03	5.00	3.46
Paper	68	0.14	0.83	0.03	0.66	3.71	4.99	2.20
Valuable	55	0.77	0.21	0.02	0.94	4.87	4.69	3.03
Green	54	0.14	0.80	0.06	0.61	3.38	5.07	1.87
Easy to carry	37	0.16	0.80	0.03	0.77	3.89	5.07	2.91
Durable	26	0.22	0.67	0.11	0.54	3.83	4.72	2.07
Divisibility	16	0.44	0.43	0.14	0.83	4.45	4.88	2.68
Ink	15	0.15	0.83	0.02	0.63	3.61	5.12	2.36
Tangible	14	0.17	0.63	0.20	0.72	4.03	5.09	2.85
Depicts president	14	0.21	0.71	0.08	0.67	3.99	5.23	2.51
Use	12	0.78	0.19	0.03	0.94	5.05	4.77	3.35
Soft, flexible	11	0.11	0.83	0.06	0.60	3.52	5.05	1.94
Dirty	11	0.23	0.13	0.64	0.48	3.73	2.66	1.82
Unit of account	10	0.70	0.26	0.03	0.92	4.94	4.48	3.00
Necessary	10	0.82	0.14	0.05	0.87	4.92	4.83	2.89
Uniform	9	0.20	0.76	0.05	0.69	3.56	5.25	2.21
Rectangular	9	0.10	0.86	0.03	0.57	3.93	5.03	2.26
Limited supply	8	0.52	0.34	0.14	0.78	4.49	4.32	2.49
Has a serial number.	6	0.29	0.68	0.03	0.75	4.13	5.12	2.73
Stable	5	0.71	0.18	0.10	0.69	4.48	4.06	1.75

Note. Properties are coded according to three empirically derived clusters: social-functional properties in dark shade, material properties in light shade. The property 'dirty' is an outlier

Indeed, participants generated material properties 37.70% more often than social-functional properties. Material properties were explained by the *creator's original intentions*.

Together, these data help to explain why participants always privileged current social status in their category judgments of money but why intended purpose also contributed to money category judgments for many people. Both a community's current rules/practices and the creator's original intentions help make sense of the properties people bring to mind when thinking about money, and so both play a role in how people categorize money. But, despite the explanatory relevance of the creator's original intentions, the community's current rules and practices better explain the social-functional properties of money, which are more fundamental to its kindhood.

5.6 General Discussion

The experiments discussed in this paper begin to shed light on the way in which people conceptualize money. Across four studies, we found that participants represent money through two major clusters of properties: social-functional properties (e.g., value) and material properties (e.g., depicting Presidents). Participants judged that social-functional properties were more conceptually significant than material properties; the social-functional properties are core to money as a kind, rather than merely statistically associated with money. Nevertheless, the material properties of money often came to mind. Participants judged that social-functional properties were explained by the current social position of the object, whereas material properties were best explained by the kind of object they were intended to be by their creator. The current social position an object occupies explains its social functional properties and so is causally central to money's core properties; nevertheless, defunct currency objects persist as a kind of borderline case because the reason they exist and have their material properties is because they were created to be money. Overall, our results suggest that people represent money as involving non-obvious causal processes, and that these causal processes guide their category judgments.

The first two studies indicate that people categorize objects as money when they currently occupy that position in society. But objects that were originally created to occupy that position (i.e., defunct currency objects) were still money to some participants. Between the two, social position was consistently important to most participants in all designs, and therefore the best description of people's representations of money is as a position in society that objects can occupy rather than a special kind of physical object. In contrast, we found a bimodal pattern of responding with defunct currency objects: participants either strongly agreed or disagreed that defunct currency objects were money. We think this reflects the conflicting intuitions that defunct currency objects are not money (since they do not currently occupy that position in society) but also have to be money (since they exist because they were made to be money).

These findings complement and help to extend past accounts of money. Prior work found that people believe that acting with currency objects (i.e., purchasing goods with dollar bills) depends on the current position those objects occupy in society, whereas acting with standard artifacts depends on the physical structure of the object (Noyes et al., 2018). Consistent with this, we found that category judgments of money reflect social structure more than physical structure, such that an object can be entirely money or not money depending on its current position in society. Slightly qualifying this finding, however, the material properties associated with currency objects contribute to how people represent money, and so the intended purpose of objects contribute to money somewhat independently of current social position. This opens up the possibility of some overlap in how people think of money and standard artifacts. Though, as we argue below, we think this overlap reflects a unique causal structure for the conceptualization of money.

5.6.1 Connections to Theories of Artifacts

There may be ways of broadening theories of artifacts to encompass institutional objects like money. This broadening would require isolating the conceptual overlap between institutional objects and standard artifacts from the unique causal structures of these domains. Artifact category judgments involve physical affordances, current use, and their original intended purpose (see Chaigneau et al., 2004 for overview). We believe that all three of these properties have analogs with money once transformed to account for the distinctly social (rather than physical) structure of money.

Physical affordances seem ill-suited for understanding money. A physically identical object can be entirely money or entirely not money depending on the current position it occupies in society (as the case of shells and cigarettes demonstrates). And of course, a medium of exchange can practically have no concrete instantiation—as the move toward digital forms of exchange (exemplified by Bitcoin) has made clear. But money is possibly understood through its *social affordances*, which refer to the way institutions make possible new forms of acting (Noyes, et al., 2020). This could explain why collective recognition was so important to participants' category judgments in Study 3: even though an object was not currently functioning as a medium of exchange, participants said it was money if people recognized it as money. Participants may have reasoned that such recognition made it possible for individuals to use an object to purchase goods, even though they were not doing so in practice. Our results thus contribute to recent theoretical and experimental research on the semantics of social objects, and in particular to the debate on their abstract vs concrete properties (Pexman et al., 2023). A more general theory of affordances could consider not only the relationship between physical objects and people's bodies (Gallagher & Crisafi, 2009; Petracca & Gallagher, 2020) but also the relationship between people and objects in social contexts. In general, people may represent object kinds through stable possibilities of action.

Intended purpose makes an important contribution to how people categorize standard artifacts (Bloom, 1996; Matan & Carey, 2001).[4] We found that intended purpose was not sufficient for people to categorize artifacts as money. Nevertheless, intended purpose clearly made an important contribution to category judgments about money, and therefore it may be important to both institutional objects and standard artifacts. However, this overlap has to be qualified by money's unique causal structure. When people represent the intended purpose of a standard artifact, they tend to have in mind individual creators, and often inventors and designers with substantial intellectual involvement in the object. Currency objects in contrast cannot be created by individuals. An individual that makes a dollar bill has made a counterfeit. The creation of dollar bills is restricted to socially recognized manufacturers (the Bureau of Engraving and Printing, the European Central Bank, etc.). Also, people may not have in mind intellectual involvement like invention and design; the creative act of making an object is less important than the act of making an object that fulfills certain agreed-upon rules. Therefore, we think that to encompass institutional objects like money, intentional theories of artifacts would need to be modified substantially.

The reason intended purpose contributes to category judgments in both domains may be that intended purpose explains why they were created in the first place. Even though defunct currency objects, like broken cleaning brooms, lack the right affordances and effective function, their existence cannot be explained any other way. Moreover, the fact that they have been used as money in the past may induce people to perceive them as objects with the potential or disposition to become money, should the appropriate conditions be instantiated again. These features arguably make defunct currency objects at the very least a kind of borderline or honorary member of the artifacts category.

Finally, current use may contribute to category judgments of both standard artifacts and institutional objects. As before, in the case of institutional objects current use would need to be modified, however. A private individual can use a rock to hammer a nail. But for an individual to use a shell to purchase goods, the entire community needs to coordinate their use of the object (Noyes et al., 2018). This broader sense of current use did contribute to how people categorize money: we saw this in Study 3, where people were more likely to categorize an object as money if the community was currently using it as a medium of exchange. However, as with standard artifacts, current use does not appear to be a primary driver of money category judgments.

In conclusion then, we think that (1) Institutional objects cannot be readily accounted for by theories of standard artifacts (Noyes et al., 2018), and so institutional objects need to be studied in their own right. (2) Nevertheless, studying institutional objects may promote a deeper understanding of how people represent standard artifacts, and how people represent objects in general. People may have general priorities in how they represent objects (they may ask: *Can I use this*

---

[4] Notice that Matan and Carey (2001) used a forced choice paradigm, rather than a rating scale.

to Φ? Why was this made? For what purpose?) that interact with the specific causal structure of the objects. Systematically comparing institutional objects with standard artifacts might help isolate these conceptual priorities.

5.6.2 Looking Ahead: Cashless Transactions

An interesting topic for future research is how people represent cashless transactions. Our description of people's representation of money predicts that if a community shifted entirely to cashless transactions, participants would categorize those cashless transactions as money (or perhaps the verifiable, digital records involved in those transactions as money) but that many participants would continue to categorize the defunct currency object as money too. Further, people's experiences with currency objects may alter their concepts of money: for example, people in advanced economies that have become cashless, particularly those who have spent their entire adult lives in such societies, may represent money differently than those who have or have had regular exposure to physical currency objects. An even more fundamental question is whether the ability to represent money in abstraction from currency objects (as we find here) is conceptually necessary for transitioning to a cashless society; this question could possibly be answered by sampling subjects at different time points during cognitive development (along the lines of Noyes et al., 2018).

Perhaps participants in our sample were particularly open to representing money as the current social position of an object (rather than as an object) because the United States is substantially cashless and people have participated in cashless transactions. Participants in other cultures who have not been exposed to cashless transactions may reveal different responses. Our hypothesis based on existing data is that people will represent money as a current social position without exposure to cashless transactions. Elementary-school children understand that institutional artifacts are depend on the community's rules and practices (Noyes et al., 2018), and we believe it is this causal understanding that guides people to categorize as money objects occupying that position in society. We think this causal insight does not depend on exposure to cashless transactions. Nevertheless, this remains an empirical question.

5.6.3 Conclusion

This is the first large study to systematically investigate how participants categorize money. We found that participants primarily represented money as a social institution, and thus categorized objects as money when they occupied the relevant position in society. We also found that participants often brought to mind material properties of currency objects when thinking about money. For many participants, defunct currency objects remained money even when they no longer occupied that position in society, which led to a bimodal pattern of responding to defunct currency objects. Therefore, when institutions lead to the creation of new physical objects that would not exist otherwise, these institutional products may take on prominence in how participants represent the institution, and these products may still be understood in virtue of the institution even when they no longer occupy that position in society. Even so, our findings add further support for unique cognitive signatures of institutional concepts and the need to study institutions as a distinct domain of ordinary concepts. Institutions are a significant part of people's daily lives across societies, and studying institutions can provide insight into general priorities in people's concepts.

Acknowledgements The results of the experiments reported in this paper were presented at seminars at the universities of Groningen and Lyon, at a meeting of the International Network of Economic Method, and at the Philosophy of Finance workshop held in Rome, from which this volume originates. We thank audiences at these events, as well as two anonymous reviewers who provided a number of useful comments and suggestions. During the preparation of this paper Francesco Guala was supported by two projects funded by the Italian Ministry of University and Research (MUR): the project "Normative Kinds" (PRIN 2022SYAW7A), and the project "Departments of Excellence 2023-2027" awarded to the Department of Philosophy "Piero Martinetti" of the University of Milan.

References

Ahn, W. K. (1998). Why are different features central for natural kinds and artifacts? The role of causal status in determining feature centrality. *Cognition, 69*, 135–178.

Ahn, W. K., Kim, N. S., Lassaline, M. E., & Dennis, M. J. (2000). Causal status as a determinant of feature centrality. *Cognitive Psychology, 41*, 361–416.

Alexander, J., & Weinberg, J. M. (2007). Analytic epistemology and experimental philosophy. *Philosophy Compass, 2*, 56–80.

Berinsky, A. J., Huber, G. A., & Lenz, G. S. (2012). Evaluating online labor markets for experimental research: Amazon. Com's mechanical Turk. *Political Analysis, 20*, 351–368.

Bloom, P. (1996). Intention, history, and artifact concepts. *Cognition, 60*(1), 1–29.

Burnham, M. J., Le, Y. K., & Piedmont, R. L. (2018). Who is Mturk? Personal characteristics and sample consistency of these online workers. *Mental Health, Religion & Culture, 21*, 934–944.

Celano, B. (1999). Collective intentionality, self-referentiality, and false beliefs: Some issues concerning institutional facts. *Analyse & Kritik, 21*, 237–250.

Chaigneau, S. E., Barsalou, L. W., & Sloman, S. A. (2004). Assessing the causal structure of function. *Journal of Experimental Psychology: General, 133*, 601–625.

Deen, M., & de Rooij, M. (2020). ClusterBootstrap: An R package for the analysis of hierarchical data using generalized linear models with the cluster bootstrap. *Behavior Research Methods, 52*, 572–590.

Diesendruck, G., & Haber, L. (2009). God's categories: The effect of religiosity on children's teleological and essentialist beliefs about categories. *Cognition, 110*, 100–114.

Frasser, C., & Guzmán, G. (2020). What do we call money? An appraisal of the money or non-money view. *Journal of Institutional Economics, 16*, 25–40.

Gallagher, S., & Crisafi, A. (2009). Mental institutions. *Topoi, 28*, 45–51.

Gelman, S. A. (2003). *The essential child: Origins of essentialism in everyday thought*. Oxford University Press.
Gelman, S. A. (2013). Artifacts and essentialism. *Review of Philosophy and Psychology, 4*, 449–463.
Gelman, S. A., & Rhodes, M. (2012). Two-thousand years of stasis. In K. S. Rosengren, S. Brem, E. M. Evans, & G. M. Sinatra (Eds.), *Evolution challenges: Integrating research and practice in teaching and learning about evolution* (pp. 200–207). Oxford University Press.
Guala, F. (2016). *Understanding institutions: The science and philosophy of living together*. Princeton University Press.
Huff, C., & Tingley, D. (2015). "Who are these people?" Evaluating the demographic characteristics and political preferences of MTurk survey respondents. *Research & Politics, 2*, 2.
Keil, F. C. (1989). *Concepts, kinds and development*. MIT Press.
Khalidi, M. A. (2015). Three kinds of social kinds. *Philosophy and Phenomenological Research, 90*, 96–112.
Knobe, J., & Nichols, S. (Eds). (2007). *Experimental philosophy*. Oxford University Press.
Kraus, M. W., & Keltner, D. (2013). Social class rank, essentialism, and punitive judgment. *Journal of Personality and Social Psychology, 105*, 247–261.
Lagerspetz, E. (1989). *A conventionalist theory of institutions*. Acta Philosophica Finnica.
Leslie, S. J. (2013). Essence and natural kinds: When science meets preschooler intuition. *Oxford Studies in Epistemology, 4*, 108–165.
Mäki, U. (2020). Reflections on the ontology of money. *Journal of Social Ontology, 6*, 245–263.
Malt, B. C., & Sloman, S. A. (2007). Artifact categorization: The good, the bad, and the ugly. In E. Margolis & S. Laurence (Eds.), *Creations of the mind: Theories of artifacts and their representation* (pp. 85–123). Oxford University Press.
Margolis, E. E., & Laurence, S. E. (2007). *Creations of the mind: Theories of artifacts and their representation*. Oxford University Press.
Matan, A., & Carey, S. (2001). Developmental changes within the core of artifact concepts. *Cognition, 78*, 1–26.
Medin, D. L., & Ortony, A. (1989). Psychological essentialism. In S. Vosniadou & A. Ortony (Eds.), *Similarity and analogical reasoning* (pp. 179–196). Cambridge University Press.
Norenzayan, A., Smith, E., & Kim, B. (2002). Cultural preferences for formal versus intuitive reasoning. *Cognitive Science, 26*, 653–684.
Noyes, A., & Dunham, Y. (2017). Mutual intentions as a causal framework for social groups. *Cognition, 162*, 133–142.
Noyes, A., & Dunham, Y. (2020). Groups as institutions: The use of constitutive rules to attribute group membership. *Cognition, 196*, 104143.
Noyes, A., & Keil, F. C. (2019). Generics designate kinds but not always essences. *Proceedings of the National Academy of Sciences, 116*, 20354–20359.
Noyes, A., & Keil, F. C. (2020). Collective recognition and function in concepts of institutional social groups. *Journal of Experimental Psychology: General, 149*, 1344–1359.
Noyes, A., Keil, F. C., & Dunham, Y. (2018). The emerging causal understanding of institutional objects. *Cognition, 170*, 83–87.
Petracca, E., & Gallagher, S. (2020). Economic cognitive institutions. *Journal of Institutional Economics, 16*, 747–765.
Pexman, P. M., Diveica, V., & Binney, R. J. (2023). Social semantics: The organization and grounding of abstract concepts. *Philosophical Transactions of the Royal Society B, 378*, 20210363.
Prasada, S., & Dillingham, E. M. (2006). Principled and statistical connections in common sense conception. *Cognition, 99*, 73–112.
Prasada, S., & Dillingham, E. M. (2009). Representation of principled connections: A window onto the formal aspect of common sense conception. *Cognitive Science, 33*, 401–448.
Rhodes, M. (2013). How two intuitive theories shape the development of social categorization. *Child Development Perspectives, 7*, 12–16.
Rosenberg, A. (2016). *The philosophy of social science* (5th ed.). Westview Press.

Roversi, C., Borghi, A. M., & Tummolini, L. (2013). A marriage is an artefact and not a walk that we take together: An experimental study on the categorization of artefacts. *Review of Philosophy and Psychology, 4*, 527–542.

Ruben, D. H. (1989). Realism in the social sciences. In H. Lawson & L Appignanesi (Eds.), *Dismantling truth* (pp. 58–75). Weidenfeld & Nicholson.

Searle, J. R. (1995). *The construction of social reality*. Penguin.

Searle, J. (2010). *Making the social world: The structure of human civilization*. Oxford University Press.

Simmel, G. (1978). *The philosophy of money*. Routledge.

Smit, J. P., Buekens, F., & Du Plessis, S. (2016). Cigarettes, dollars and bitcoins–an essay on the ontology of money. *Journal of Institutional Economics, 12*, 327–347.

Thomasson, A. L. (2003). Realism and human kinds. *Philosophy & Phenomenological Research, 67*, 580–609.

Tummolini, L., & Castelfranchi, C. (2006). The cognitive and behavioral mediation of institutions: Towards an account of institutional actions. *Cognitive Systems Research, 7*, 307–323.

Vasilyeva, N., Gopnik, A., & Lombrozo, T. (2018). The development of structural thinking about social categories. *Developmental Psychology, 54*, 1735–1744.

Vasilyeva, N., & Lombrozo, T. (2020). Structural thinking about social categories: Evidence from formal explanations, generics, and generalization. *Cognition, 204*, 104383.

Villani, C., D'Ascenzo, S., Borghi, A. M., Roversi, C., Benassi, M., & Lugli, L. (2022). Is justice grounded? How expertise shapes conceptual representation of institutional concepts. *Psychological Research, 86*, 2434–2450.

Chapter 6
The Epistemic Value of a Financial Millisecond: Quotes, Speed Bumps, and Market Design

Emiliano Ippoliti

Abstract In this paper I discuss the epistemic value of a quote and the effect of technological innovations on this value and on market design. The epistemic value of a quote is the agent's belief about the *correct* price of an asset. Time, even milliseconds, is an essential feature of this epistemic value: in these milliseconds you can access certain information, and get to know how the markets is moving, and adjust your order to it consequently, so they produce an advantage that can be used to profit, or to avoid losses, from price changes that happen very quickly. The epistemic value of a financial millisecond is the advantage that can be gained by being able to trade in a millisecond faster than your competitors.

'Micro-quotes' are the result of financial technology innovations. Micro-quotes are quotes with a theoretical expiration date measured in milliseconds since they can be placed and then canceled at any time. I discuss how they produce 'ghost liquidity' and a loss of epistemic values, and how a solution is a different market design. In this case a solution is the introduction of speed bumps, which try to level the playing field for traders who do not have the fastest technology by delaying the execution of orders for a short period of time, which gives slower traders a chance to react to price changes. I examine three distinct speed bump worlds (World 1: Takers-priority, World 2: Makers-priority, World 3: No-priority) and their epistemic and moral features, and I discuss how they can be purposely designed based on predetermined objectives. This process falls under the realm of 'reverse finance', wherein we begin with specific financial market goals and then create a system that can achieve those objectives more effectively.

Finally, I examine how the choice of the target and time threshold of speed bumps is not a value-neutral or technical decision. It has a significant impact on the structure and agents of financial markets, and it is epistemically and morally loaded: it determines who has access to information and who is able to profit from market movements, and also how efficient and stable financial markets are.

E. Ippoliti (✉)
Department of Philosophy, Sapienza University of Rome, Rome, Italy
e-mail: emiliano.ippoliti@uniroma1.it

E. Ippoliti et al. (eds.), *Philosophy and Finance: Ten Open Questions*, Synthese Library 504, https://doi.org/10.1007/978-3-031-96510-4_6

6.1 Introduction

Financial markets are in general, and even primarily, epistemic systems. They provide collective mechanisms for generating and evaluating knowledge about the value of assets. To do this, they convey a lot of knowledge about a multitude of different aspects of the economy, the societies, and their prospects, and express all this knowledge by individual numbers and prices. This knowledge is 'distributed': millions of investors possess distinct information and beliefs, which are aggregated and presented in a numerical manner by financial markets and that are essential for profiting of financial markets.

In this sense, knowledge is the most valuable commodity in financial markets and a precondition for operating (and profiting) on the markets. The primary goal in trading is knowledge: trades are a way for financial agents to gain knowledge from one another's, and from market trends, and this process of learning helps them develop beliefs about asset prices that are more accurate. In this respect, stock markets are also a form of social learning.

Through this process, the financial markets impound into the price what the market *(as a whole)* thinks about a company's future potential—i.e., the prevalent or collective belief (see Gilbert, 2004). As new information becomes available, these beliefs are being revised and refined.

The complexity of this knowledge is vast, making it fragile, uncertain, and fallible. Still, all this knowledge must be expressed by means of numbers and prices. This contrast between the variety, sophistication and instability of the knowledge conveyed by the financial markets on the one hand, and the precision required by individual numbers and prices on the other hand, deeply characterizes the financial markets and makes their epistemic task so difficult, but still so useful and interesting. In addition, this explains why market design is so important: the way a market is setup, and the rules used to trade and form prices and quotes, can foster, or hinder this fundamental epistemic function of the financial systems.

Of course, stock markets are not perfect epistemic systems, since knowledge is not distributed equally among financial agents or it can be intentionally concealed: market manipulation is always a possibility, and there will always be some investors with more knowledge than others. As a result, the prices of stocks are not always reliable epistemic entities (e.g. that do not reflect the 'true' value of assets) and it is difficult and risky to infer beliefs from prices (see Thicke, 2018).

In this paper, I will examine a crucial aspect of this epistemic function of financial markets by focussing on the notion of quote in stock markets, its epistemic value and how new technologies affect it. In greater detail, I will examine the epistemic value of a quote and what is the effect of technological innovations on this value and on market design. First, I introduce the notion of epistemic value of a quote: it is defined as the agent's belief about the correct price of an asset (Sect. 6.1). This centrality of epistemic value of quotes explains why market are often designed to encourages agents to reveal their underlying beliefs (Sect. 6.1.1). I analyse how time, even milliseconds, is an essential feature of this epistemic value (Sect. 6.1.2). In fact, during these milliseconds you can access certain information, and get to know how the markets is moving and adjust your order to it consequently. Thus, even tiny fractions of time provide knowledge producing an advantage that can be used to profit, or to avoid losses, from price changes that happen very quickly. The epistemic value of a financial millisecond is the advantage that can be gained by being able to know and trade in a millisecond faster than your competitors. I then consider the 'micro-quotes', quotes that are the result of financial technology innovations that enable (algo) trader to place and cancel orders in terms of milliseconds.

Then I discuss how micro-quotes produce epistemic opacity and how this loss of epistemic values is counteracted by a different markets design, that is, introducing 'speed bumps' (§ 6.1.2). Market design is the process of deciding how financial markets should work. This includes decisions about how orders are processed or how prices are displayed (in this case how speed bumps are used). In essence, speed bumps try to level the playing field for traders who do not have the fastest technology by delaying the execution of orders for a short period of time, which gives slower traders a chance to react to price changes. In addition, I examine three distinct speed bump worlds (World 1: Takers-priority, World 2: Makers-priority, World 3: No-priority) and their epistemic and moral features, and I discuss how they can be purposely designed based on predetermined objectives (§ 6.1.3). This process falls under the realm of 'reverse finance' (see Ippoliti, 2020), wherein we begin with specific financial market goals and then create a system that can achieve those objectives more effectively.[1]

Finally (§ 6.1.4), I examine how the choice of the target and time threshold of speed bumps is not a value-neutral or technical decision. It has a significant impact on the structure and agents of financial markets, and it is epistemically and morally loaded: it determines who has access to information and who is able to profit from market movements, and also how efficient and stable financial markets are.

6.1.1 The Epistemic Value of (Micro-)Quotes

A quote is typically defined as the most recent price at which an asset traded: it is the price at which a buyer and a seller came to an agreement and at which a portion of the asset was exchanged.

The 'bid' price represents the most recent price and share amount at which a share of the asset may be purchased (i.e., buy price). The bid quote displays the price and quantity at which an existing bidder is willing to buy the shares. The 'ask' price indicates how much a current trader is prepared to take in exchange for their shares (i.e., sell price). Quotes are epistemic objects: they bring and encapsule al lot

[1] For more detail on the notion reverse reasoning, see Chap. 5 in Gabbay and Woods (2005) and Chap. 2 in Magnani (2009).

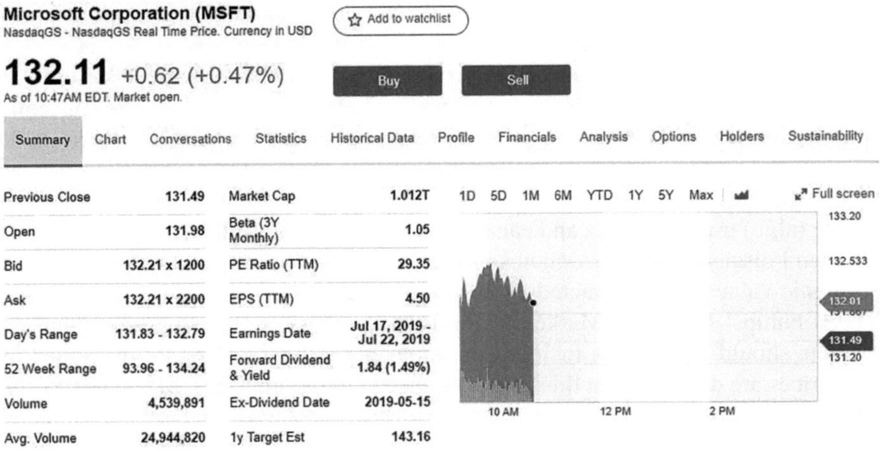

Fig. 6.1 Example of a Quote

of knowledge not only about financial 'facts', but also, and especially according to certain theories, about the financial agents and their beliefs (see Fig. 6.1).

The epistemic value of a quote can be defined, in an idealised extreme case, as the agent's belief about the *correct* price of an asset. This definition raises at least two important issues:

(i) As anticipated above, the price of actual quotes often does not carry perfect information about the agent's beliefs. As a matter of fact, the price can be the outcome of opportunistic or strategic behaviour. e.g., agents initially underbid in order to test if the other party has a different belief. This behaviour can deteriorate a quote's epistemic value, since this does not convey information about what the agents really believe, but rather what they want the other agents to believe. In order to uphold the epistemic value of quotes and markets, as elaborated in the subsequent sections (Sect. 6.1.1), markets are frequently structured with deliberate mechanisms to encourage participants to disclose their fundamental beliefs. This, in turn, underscores the significance of a quote in conveying (albeit imprecise) information about an agent's belief.
(ii) The interpretation of the term "correct" price can vary depending on one's theoretical stance on pricing. We can identify at least two primary perspectives on this matter. Some adhere to a "realist" standpoint by asserting the existence of a "correct" (or "real") price—an inherent value of the asset,[2] as proposed by the efficient market hypothesis (EMH, see Fama, 1970; Fama, 1991; Samuelson, 1965). This perspective supports a *foundational* approach to pricing, where

[2] On the controversial notion of 'true/intrinsic value' of financial assets see also de Bruin et al. (2020).

> prices are seen as a source of true knowledge (justified beliefs). This is because prices can be traced back to something secure and objective, which serves as the fundamental basis from which all other knowledge and beliefs stem.

Others maintain a 'constructive' or even an 'anti-realist' standpoint by arguing that a price is simply a mutually satisfactory agreement, it is socially constructed—e.g., the reflexive market hypothesis (see Flanagan, 1981; Popper, 1957; Soros, 1987) and behavioural finance (Kahneman et al., 1982)—, and hence it can be 'wrong' in many ways. These approaches endorse a *fallibilist* view of prices, that is, the idea that prices provide us with knowledge that are invariably vulnerable, which cannot be traced back to something assured and objective and then may turn out to be false. This view does not hold that knowledge is unavailable here, but rather that it is always provisional.

In more detail, the efficient market hypothesis maintains that there is such a thing as the intrinsic value for an asset, which is given by the fundamentals of this asset[3] and to which the quote of the asset can be traced back, and hence its price is 'right' most of the time, in the sense that the dynamics of asks and bids converges quickly towards this right price, that is, the one reflecting all the relevant information about that asset. Since the stock quote reflects all available information, it is almost impossible to consistently outperform the market. The well-known EMH's explanation for this goes as follows: if all investors have access to the same information, then they will react in the same way to new information and they will all reach the same conclusion about the value of a stock, and the stock price will reflect that value.[4] As a result, the EMH states that stock market quotes do not provide any new information that can be used to make better investment decisions. Thus, information cannot be used to gain an edge over the market.

On the contrary, the RMH argues that the price is almost always 'wrong', in the sense that bids and asks do not converge towards the 'right' price because agent's biases influence the value of quotes in a way that may push it away from what it is supposed to be its 'fundamental' or 'real' value, if any. However, the RMH argues that stock prices can be self-fulfilling prophecies. In brief, this means that if investors believe that a stock price is going to go up, they may be more likely to buy it, which will drive the price up (and change the fundamentals themselves of a company in the process). This suggests that stock market quotes can have some epistemic value, as they can be used to size investor sentiment and trends. This fallibilist view of prices also characterizes the behaviour finance (BF), which argues that investors are not always 'rational', and that their emotions can influence their investment decisions.

---

[3] The typical fundamentals are earnings (a measure of its profitability); cash flow (the amount of money that a company generates from its operations); dividends (the payments that a company makes to its shareholders out of its profits); debt-to-equity ratio (a measure of a company's financial leverage); industry growth.

[4] Notably, the three versions of the EHM (weak, semi-weak, and strong) make different statements about the flow of information in financial markets.

Consequently, these approaches confer different epistemic values to quotes:

(i) the EMH argues that quotes reflect all available information, making it impossible to consistently outperform the market. The epistemic value of quotes is maximum most of the times: the price provides everything that we need to know (!),[5] and it expresses the prevailing belief, the trends, and the fundamentals all at once (there is nothing left to know about the market). There is no substantial difference between the belief of a correct price and the correct price itself, since they became the same in a tiny amount of time.
(ii) RMH and BF state that market quotes can have epistemic value since they can be used to gauge investor sentiment and trends, which will determine the 'right' price because of the reflexive dynamics, when it is at work. The price, then, has not that high or maximum epistemic value as argued by the EMH.

Now, one key and very general function of the financial markets is to facilitate mutually beneficial solutions to coordination problems. In this sense, the epistemic value of a quote is essential to fulfilling at least three core functions of financial markets (see e.g., Chami et al., 2010; Merton & Bodie, 1995),[6] such as discovering financial prices, providing liquidity, dealing with information problems. This centrality of epistemic value of quotes explains why market are often designed to encourages agents to reveal their underlying beliefs. I will give some examples of these mechanisms.

In a broad sense, most of the markets may be thought of as epistemic processors, namely, a set of mechanisms and structures that try to get, aggregate, and compose the epistemic state (see Ippoliti, 2023, 2024, 2024a) of millions of investors, each with a single set of information and beliefs.

More specially, many financial devices can be thought of as tools to reveal this epistemic state of the investors.

Traditional devices such as 'futures' and 'options' can be seen as implicit incentives to reveal agents' beliefs about the current and future states of the markets—even if at different degrees. In fact, the belief is stronger for futures than options since in the former case you must sell or buy at the expiration date.

A more explicit example is the way a market is designed. For instance, the distinction between lit markets and dark markets (dark pools) can be understood just in terms of potential rewards for revealing beliefs. In greater detail, in a lit market the traders do reveal, and can see and get, a lot of information such as the best 'bid', the best 'ask', the five best bids, the five best asks, the quantities at those prices, and total quantity on the bid side, the total quantity on the ask side, and details of previous prices and transactions. On the other hand, in a dark pool, all

[5] See Williams and Dobelman (2020) for an extensive critique of this tenet.

[6] The main basic functions of the financial systems are: 1. Transferring resources across time and space; 2. Pooling resources and sharing ownership; 3. Discovering financial prices; 4. Providing liquidity; 5. Dealing with information problems; 6. Clearing and settling payments; 7. Managing risks.

the traders can know is the price—and a common way to determine the price in a dark pool is to take the midpoint of an equivalent lit market. But a trader doesn't know crucial information such as whether the execution would be instantaneous, or whether there'll be entering a queue. So, traders must reveal, but also see, a very different, and limited, set of information.

Now, lit markets, which display much more information, present the so-called 'order protection rule': when an exchange has the best price for a stock at any given moment, orders must be routed its way. Lightly regulated alternative platforms, like dark pool, don't have that same advantage. That is, exchanges are encouraged to attract orders with a lot of information and agents' beliefs attached to them and are protected and rewarded if they do this. Nevertheless, this protection of the epistemic value of quotes and markets is compromised through the utilization of emerging technologies.

6.1.2 The Epistemic Opacity of Micro-Quotes

A new phenomenon has emerged with the development and increasing use of high-speed trading algorithms (e.g. HFT) in financial markets, whereby large volumes of orders are placed automatically at very high speeds across different venues. In more detail, it is possible to place and then cancel large orders at the speed of light before exercise, that is, to place large orders and then cancel them in term of milliseconds before their execution, and to move them across several exchanges.

This can be done only by machines on certain infrastructures (at least at this speed). This new phenomenon has generated new market's dynamics, in particular a form of 'phantom, or ghost, liquidity' (GL), which have been analysed and popularized extensively (see e.g. Levine, 2015, 2016, 2019, 2019a) and even fictionalized (see Michael Lewis's book "Flash Boys"). GL is particularly interesting from an epistemic viewpoint because it affects and dilutes the epistemic values of quotes. Let's see how.

Here is a first simple example of market's dynamics producing ghost liquidity. The stock s is trading at € 10. I place an order to buy it at € 9.99 and my order goes onto the order book. I am bound by it: if the value of s goes down to € 9.95, someone will take my bid, and I must buy the stock at € 9.99 (instead of the better price € 9.95). On the contrary, if the value of the stock s goes up to € 10.05, I will not get any stock. This is a typical situation where the risk is 'asymmetric': I buy the stock when it is going down, but I don't buy it when it is going up. Fortunately, there is a way of managing this risk. This is how: if, *somehow*, I get to *know* in advance that the market is moving against me and if I am the fastest to cancel my order at € 9.99 (since orders can be cancelled at any time before their execution), then I can avoid buying s at € 9.99, since I know that its true value is € 9.95. In this case, I could update my price by submitting a new buy order at € 9.94 and I could try to get a better price. This is precisely what high-frequency trading enables to do.

Table 6.1 An example of how HFT works

Venue A	Venue B
t at € 10	*t* at € 10
⇩	
t at € 10.01	
	HFT buys *t* at € 10 at time *h*
HFT sells *t* at € 10.01 a time $h + m$ (makes a profit of € 0.01 in time *m*)	⇦
	t at € 10.01

Here is a second example of a market's dynamics producing GL. Let us suppose that I am a market maker ready to buy or sell the stock. In this case, the situation is it's even more difficult since the risk is 'symmetric', that is, in both directions. In fact, in this case when the stock *s* is at € 10, not only do I market maker submit a bid to buy it at € 9.99, but I also submit an ask (offer) to sell it at € 10.01. Thus, if the stock quote goes up to € 10.05, I will have to sell it at the worse price € 10.01. If the stock goes down to € 9.95 I will have to buy it at the worse price € 9.99. But now, again, there is a way of managing this risk: if I am the fastest to cancel my orders (since orders can be cancelled at any time before their execution) I can avoid buying *s* at € 9.99 when its current value is € 9.95, or selling *s* at € 10.01 when its current value is € 10.05.

This behaviour forms the foundation of the high-frequency trading (HFT). In fact, a typical high-frequency trading strategy operates as follows:

At a given moment, an asset "t" is available for purchase at the same price on various exchanges simultaneously (e.g. € 10)

The price of "t" changes on one exchange, let's say exchange A, due to market activity—for example, the price rises because someone buys the asset (e.g. € 10.01).

You take advantage of this price difference by swiftly purchasing the asset on another exchange, like exchange B, a fraction of a second later before it changes in B, and then promptly reselling it at the higher price on exchange A, thereby generating a profit (see Table 6.1).

Of course, you must be super-fast in order to exploit this process successfully: the price could move against you, when you buy the asset on the venue B and sell it at the venue A (e.g. the price might swiftly drop to €9.99 within a few milliseconds, rendering your offer at €10.01 obsolete), leaving you with an asset that you acquired solely for the intention of reselling it within a few milliseconds, rather than holding it based on a genuine belief in its value. Thus, HFT enacts well-known trading strategies (such as market making or arbitrage) in a technologically advanced fashion by exploiting the features of the market infrastructures (see e.g. Mackintosh, 2019; Mackintosh, 2020). It exploits very-short-lived informational asymmetries and advantages. The epistemic effect of HFT is clear: a reduction of the commitment of algo traders (in particular market makers) on their submitted orders.

This means that these traders are less likely to follow through on their orders, which dilutes the epistemic value of quotes in the process.

More specifically, we can identify at least two types of 'ghost liquidity':

(1) The first type of GL occurs when HFT traders place duplicate limit orders on competing venues, intending for only one of the orders to execute, and when one does execute, duplicates are cancelled. Thus, in this situation, orders are cancelled conditional on an order submitted by the same trader being filled on another venue.
(2) The second type of GL occurs when orders are explicitly intended not to be executed (under any circumstance). In the category falls activities like spoofing or pinging. In this situation, cancellations of orders are a sign of a potential illegal activity and hence are very risky to perform.

It is important to note that this practice is not forbidden or illegal. As a matter of fact, to place and then cancel orders before their execution, even in milliseconds, can be done for several legal and good reasons. Typical cases of good reasons are at least four:

1. To update our prices to reflect changing conditions.
2. To be a proper market maker, which requires moving prices continually and cancelling a lot of orders to respond to changing supply and demand.
3. To use different type of orders. Certain orders would only execute if they can fully complete. If the order cannot be completed, it would be immediately cancelled. Typical examples are fill-or-kill (FOK) orders or immediate-or-cancel (IOC) orders. A FOK order is an order to buy or sell a security that must be filled in its entirety or canceled entirely. This type of order is used by investors who want to ensure that they get the entire amount of the security they want at the price they want, or not at all. For example, an investor might use a FOK order to buy 1.000 shares of an illiquid stock. If the investor is only willing to buy all 1.000 shares at a specific price, and the market is not liquid enough to fill the entire order at that price, the FOK order will be canceled. An IOC order is canceled if it is not filled immediately. FOK and IOC orders are often used in conjunction.
4. Moreover, orders cancellation is accepted by exchanges since the presence of massive quantity of orders enables the exchanges fulfilling several of their core functions—like providing liquidity. Thus, profusion and multiplication of quotes is encouraged by many venues. Essentially, if a trading venue manages to draw numerous investors willing to buy stocks at €9.99 and also attracts many investors offering to sell stocks at €10.01, then any individual seeking to make a significant purchase (or sale) can readily and immediately do so at €10.01 (€9.99) respectively. Consequently, the stock exchange effectively fulfills a crucial function: providing liquidity by enabling actual investors to execute their trades instantly. As a result, trading venues have a vested interest in attracting market makers and generating a high volume of orders.

This practice and the consequent GL give rise to several epistemic issues, all stemming from a fundamental fact, namely, the awareness that orders can be canceled at any moment before execution, particularly when there's an indication that the price might move unfavourably (even before others become aware of the shift). As a result, these quotes essentially possess a theoretical expiration date measured in milliseconds. Let's call them as 'micro-quotes.' During these brief milliseconds, such quotes lack substantial epistemic value as they fail to reflect genuine beliefs of investors. Asks and bids emerge and vanish within this short timeframe, causing disruptions in the quote process. It's akin to attempting to count stars in the sky during an intense meteor shower night: the count continuously gets interrupted by the presence of meteors, making it challenging to ascertain an accurate tally of the stars.

An essential question arises here about the epistemic value of *financial time*: can the flow of time resolve this issue? Does the mere passage of time imbue quotes with greater epistemic value? In other words, should waiting a while after the expiration of these short-lived micro-quotes suffice to regain relevant information and restore their epistemic value? Unfortunately, even with the passage of time, the problem remains unsolved if orders persistently flood the market, being placed & cancelled in quick succession. This constant flow of orders being entered and withdrawn continues to impact the price, making it difficult for the quotes to (re)gain their epistemic value.

Therefore, the initial question that arises when allowing the existence of micro-quotes is whether they compromise the fundamental functions of the financial systems. In reality, several anticipated and observed outcomes related to micro-quotes seem to affirmatively respond to this question.

A key concern lies in the diminished likelihood of mutually beneficial outcomes. The ability of micro-quotes to permit very small price increments even without transaction occurring, makes it challenging for other traders to find a price that serves their interests as well. In the most extreme cases, the agent utilizing a micro-quote might capture all the benefits, leaving nothing for the opposing party.

A second concern is that micro-quotes worsen price discovery, liquidity, and the ability to deal with information problems. Price discovery is the process of determining the (fair) price of an asset. Micro-quotes make price discovery more difficult because they create a large number of small price increments, which makes it difficult to see the overall trend in prices. Liquidity is the ability to buy and sell an asset quickly and easily. Micro-quotes make liquidity worse because they make it more difficult to find buyers and sellers at a given price. Information problems arise when one party to a transaction has more information than the other party. Micro-quotes make information problems worse because they make it easier for one party to take advantage of the other party's lack of information.

Finally, micro-quotes indirectly undermine other functions of financial markets that are tied to the previous ones. For example, micro-quotes make it more difficult for investors to hedge their risks, which can lead to increased volatility in the markets. Additionally, micro-quotes can make it more difficult for market makers to operate, which can lead to decreased liquidity in the markets.

Moreover, 'micro-quotes' raise philosophical issues. The very first one is an ontological question, that is, about the nature of a quote see (O'Hara, 2010). In the case of micro-quotes, a quote does not express the match between actual supply and demand, since the price of an asset changes even if no transactions occur (see Ippoliti, 2020a; Ippoliti, 2021a). This raises the question of what a quote is, if it no longer reflects the underlying market forces.

Other philosophical issues raised by micro-quotes include the following:

- Do micro-quotes reduce the epistemic value of a quote *to zero or almost*? In fact, if micro-quotes do not reflect the underlying market forces, then they may have little or no epistemic value. This leads to another important issue, that is:
- Do micro-quotes produce *epistemic uncertainty*? Epistemic uncertainty refers to the lack of knowledge. Micro-quotes may increase epistemic uncertainty by making it more difficult to determine the value of an asset.
- Do micro-quotes produce epistemic injustice? Epistemic injustice refers to the unfair distribution of epistemic goods, such as knowledge and understanding. Micro-quoting may produce epistemic injustice by disadvantaging those who do not have the resources to track the constantly changing prices or to trade at the light speed.
- Do micro-quotes make markets less efficient? In effect, they can lead to price volatility and price distortion, which can make it difficult for investors to get the best price for their orders. According the EMH, the price of a security reflects all relevant information. However, micro-quotes can interfere with this process by introducing noise and uncertainty into the market. This is because micro-quotes do not carry any valuable information, and they can actually deteriorate the quality of the information that is already in the market. Thus, if micro-quotes do interfere with market efficiency, it would make it possible to consistently outperform the market. This is because investors who are able to trade quickly and take advantage of micro-quotes would have an advantage over other investors.[7]

Now, we can answer 'yes' to all these questions and look for solutions (§ 2). But you can also provide a more complex answer (§ 3). Let us see how.

6.1.3 Market-Based Solution to Micro-Quotes: Speed Bumps

There are several possible solutions to the problems caused by micro-quotes. One solution is governmental. This would involve institutions like the SEC introducing policies, rules, or laws to limit or eliminate micro-quotes. However, this solution has not yet been adopted and it is difficult to advance because of the good reasons for placing and cancelling orders in milliseconds, as we have seen. Another solution

[7] However, there is some debate about whether micro-quotes actually make markets less efficient. Some people argue that micro-quotes actually improve market efficiency by increasing liquidity and reducing transaction costs (see e.g. Harris, 2008).

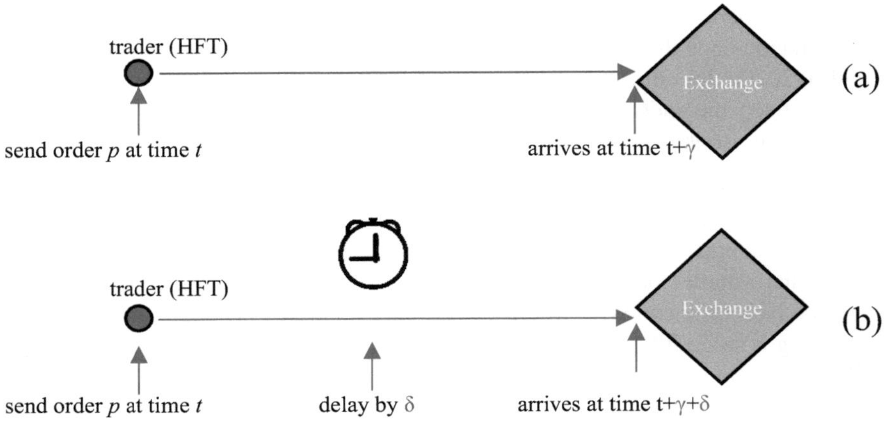

Fig. 6.2 How speed bumps work

is market-based. In the following, I will examine the main market-based solution, which is the 'speed bump'. A speed bump is a very short, intentional[8] delay that an exchange applies to the processing of incoming orders. This delay can be used to alter the 'natural' flow of orders, allowing some order actions to move faster than others (see Fig. 6.2).

There are two types of speeds bumps:

(i) **Symmetric**, whereby the delay is imposed equally to all market participants and all orders. It does not change the natural flow of orders, and hence it is a not an interesting solution in our case.
(ii) **Asymmetric**, whereby the delay is not imposed equally on certain exchange-designated market participants or on certain, but not to all, order types. For example: an exchange might delay all orders except cancellations. Consequently, a cancellation received slightly after an order of a different type might be processed first, while venues with conventional limit order books would process orders strictly in the time sequence they are received. By 2020, more than a dozen markets in stocks, futures, and currencies from Toronto to New York to Moscow slow trading via speed bumps or similar features. Also, different venues set different speed bump lengths or design (see e.g. Table 6.2).

To illustrate how an asymmetric speed bump work, let us consider an example. Let us suppose that a stock is quoted at € 10 in an exchange venue, and that a market maker has a bid of € 9.99 and an offer of € 10.01 for this stock. Now, let us suppose

[8] It is important to note that other exchanges also experience delays between the receipt and execution of orders. This is due to the physical limitations of the internet and the way in which orders are processed. These delays can be similar or even longer than the speed bumps that are intentionally introduced by some exchanges. However, the key difference is that these delays are not intentional. They are simply a by product of the way that the internet, or communication, works.

Table 6.2 Some venues with different designs of speed bumps

Exchange	Target of delay	Length of delay (in milliseconds)
IEX	All bet pegged orders	350
Thomson Reuters	Non-cancellations	0–3
Aequitas NEO	Liquidity takers	3–9
TSX Alpha	Liquidity takers	1–3
Eurex exchange	Liquidity takers	1 or 3

that there is some news—e.g. a trade on another exchange, or an earnings release, etc.—that push the price of the stock down to €9.95. Here enter the speed bumps: let us suppose that the exchange venue delays liquidity-taking orders, but doesn't delay market makers' ability to modify their existing limit orders. Then the market maker gets an extra split second to cancel her orders and change her bid and offer to € 9.94/ € 9.96 before anyone else has a chance to buy from her at € 9.99.

The effect of an asymmetric speed bump is clear: since orders can be cancelled at any time, the market maker is likely to be the fastest to update her prices and so she can avoid buying stock at € 9.99 when its value falls to € 9.95 by cancelling her quotes at € 9.99 and submitting new orders (bid/ask) at € 9.44/ € 9.46.

In this regard, speed bumps have been compared to the 'last look' function found in certain trading systems, such as electronic platforms used for foreign exchange trading (FX) (see e.g. Levine, 2015). The last look feature enables market makers to withdraw their orders briefly after they have been matched. The 'last look' feature has stirred controversy and even been deemed illegal on certain exchanges due to its potential for allowing market makers to evade their commitments, raising questions about its ethical implications (since it is considered cheating).

Still, there is a substantial difference between the last look and the speed bump.

Regarding the last look, liquidity providers respond only after receiving the order, granting them time to assess the market. Consequently, market makers gain an informational advantage *post-order*, due to the last look feature. On the other hand, the speed bump aims to eliminate the informational advantage held by faster traders by briefly pausing certain orders, allowing some of the information asymmetry to dissipate. As a result, liquidity providers must make their decisions independently of any incoming order flow and need to act swiftly to capitalize on it.

In these cases, every millisecond counts, as certain informational advantages dissipate and other don't in such a tiny amount of time.

It is not surprising that speed bumps have been a pop topic recently (see e.g. Aoyagi, 2020; Gonçalves et al., 2019; Khapko & Zoican, 2020) and that the main focus is on *asymmetric* speed bumps. Asymmetric speed bumps typically slow incoming "take" orders by a few milliseconds, effectively giving counterparties extra time to cancel or modify their existing orders. The benefits of speed bumps for liquidity providers are clear: they gain an enhanced chance to adapt their prices in rapidly changing markets and protect themselves from being disadvantaged by faster traders (or even 'adversely selected'). However, the overall impact on other traders is less clear, as the widespread use of speed bumps can also raise concerns. It creates uncertainty for investors since they cannot be certain about the prices they observe on the exchange.

Thus, on the one hand, the introduction of the speed bump aims to counter the loss of epistemic value caused by ghost liquidity and high-frequency trading (HFT). However, on the other hand, it can result in epistemic uncertainty and lead to a further loss of epistemic value.

The introduction of speed bumps raises further epistemic concerns. In fact, by setting a speed bump, a venue becomes 'dark' for a brief period of time, meaning that it hides information (similarly to a dark pool) for a fraction of a second. This phenomenon can influence market dynamics through the generation of information asymmetries, where certain traders possess exclusive access to information not available to others. Consequently, this discrepancy complicates the assessment of an asset's value, resulting in suboptimal trading choices and heightened risk for traders.

The choice not only of implementing speed bumps but also of their time threshold (orders length) are not neutral: they are value-laden choices because they have significant impact on the preferred structure and agents of financial markets, with epistemic and even ethical implications. In fact, the basic epistemic and ethical issues in high-frequency trading and speed bumps are:

(1) Should some orders take precedence over others?
(2) How long should an order last to be epistemically valuable? (Epistemic value of time)

To answer these questions, we must consider the two basic types of high-frequency traders:

(a) liquidity-providers, also known as market makers, are traders who create markets by placing buy and sell orders at specific prices. They play a crucial role in ensuring there is always someone willing to buy or sell a security, even in less liquid markets.
(b) liquidity-takers, that is the traders who take advantage of the liquidity provided by liquidity providers. They often utilize high-frequency trading strategies to exploit minor price discrepancies between different markets. In turn, there are two types of liquidity-takers:

(b1) 'informed' high frequency trading takers, who attempt to hit a quote prior to an impending price change.
(b2) 'uninformed' traders, who are taking liquidity at a point in time simply because they need to trade and not because of some informational advantage. For uninformed traders, a speed bump lasting a few milliseconds normally has little effect on their execution quality.

These two types of high-frequency traders serve different functions in the market. Liquidity providers contribute to making markets more efficient, whereas liquidity takers help to provide liquidity. The impact of high-frequency trading on the market depends on the balance between these two types of traders.

Tellingly, the idea of the speed bump came from IEX—the Investors Exchange (portraited in the book *Flash Boys*)—which employed speed bumps to increase the value of options offered by market makers by making it harder for them to cancel their orders when the market moves against them. These first speed bump slowed down the market makers' cancellations of orders but did not slow down the liquidity-taking orders that real investors used to exercise options against the market makers. This choice was contested, and IEX eventually had to abandon some of its speed bump characteristics to be approved as an exchange. The interesting feature of this scenario is that the speed bumps are introduced to favour the liquidity-takers (the *flash boys*), which are considered "the good" traders, while the electronic market makers who cancel orders are considered the "bad" traders.

Hence, speed bumps are meant as a form of protection for preferred financial agents and structures. There are three different speed bump designs, or "speed bump worlds," that favour different agents and structures:

World 1: Takers-priority
In this design, there is a halt period between when a price changes in one place and when it changes elsewhere. This gives takers, who are people who buy assets, a chance to react to price changes before makers can update their prices.

World 2: Makers-priority
In this design, you cannot buy an asset at the old price on other exchanges. This gives makers a chance to update their prices before takers can buy the asset at the old price.

World 3: No-priority
In this design, quotes must be processed in the order in which they come in. This means that takers and makers have an equal chance of getting their orders executed and the venue allows the market to determine which one has priority. In simple terms, if you can get your order to the other exchange before market makers have a chance to update their prices, you will be able to buy the stock at the old price. However, if market makers have already updated their prices by the time you send in your order, you will have to buy the stock at the new price.

These three designs are not only epistemically but also ethically loaded, because they attribute a special epistemic and moral status to certain traders or orders, thus producing very different epistemic and ethical structures. These various designs have distinct implications for the market. World-1 favours liquidity takers, while World-2 is more beneficial to market makers. World 3 is intended to be a neutral design. In greater detail, both World-1 and World-2 involve speed bumps and aim to address the issue of high-frequency traders attempting to exploit slower trader, who are also considered the more honest traders. However, their approaches are contradictory.

In World-1, market makers are seen as malicious high-frequency traders who exploit slower traders by updating their prices before they can buy at the original price: they are seen as predatory traders who use their speed to take advantage of slower traders. Conversely, in World-2, liquidity takers are seen as malicious high-frequency traders who exploit slower market makers by buying at the original price before they can update their prices: they are seen as predatory traders who use their speed to take advantage of slower market makers.

World-1 and World-2 show that a speed bump is not simply a form of protection for slower, traditional investors against the malicious HFTs, but it is rather the choice of protecting one sort of high-frequency traders (electronic market makers) from another (liquidity-taking HFTs), or vice versa.

In this sense the speed bumps can introduce 'epistemic injustice' from the outside, and the Word-3 (no-priority) is a way of avoiding this artificial introduction. If no one gets a halt period and no order gets a special epistemic and moral status, it is the market that determines who has an advantage the consequent epistemic and moral structure.

World 3 has two strengths:

(i) It doesn't require us to decide which orders are morally and epistemically good or bad. This could be seen as fairer.
(ii) It is less complex from an administrative standpoint.

However, World-3 also has drawbacks. First, it doesn't question the social and economic usefulness of high-frequency trading (HFT). High-frequency trading (HFT) can lead to traders moving their orders from one exchange to another only in search of the best price. However, this activity often does not create any real social or economic value. Second, HFT can lead to unnecessary investments in technology just to improve the speed of order execution. This can include investing in faster computers, better telecommunications infrastructure, and even locating trading servers closer to exchanges (*co-location*). This activity rewards those who win the technological competition, who are often the wealthiest investors. So, even World 3 is not values-free: it still favors those who have the resources to invest in the latest technology. In this way, they can strength their market power and, because of the technological costs, block new entrants to compete (see e.g. Sánchez Serrano, 2020).

Significantly, these three distinct speed bump worlds can be purposely designed based on predetermined objectives. This process falls under the realm of 'reverse finance' (see Ippoliti, 2020, 2021b), wherein we begin with specific financial market goals and then create a system that can achieve those objectives more effectively. This is accomplished by strategically modifying and balancing certain elements of the system.

To achieve this goal, it is necessary to resolve the fundamental tension in micro-quotes, in particular the tension between the quantity and the quality of quotes. In fact, on the one hand, exchanges compete to get more and more quotes (quantity). On the other hand, the proliferation of micro-quotes makes them less epistemically valuable (quality). In other words, if you have a lot of quotes, but they are all of low epistemic quality, then you have not accomplished anything. If an exchange has a lot of standing limit orders, but market makers can cancel those orders whenever the market moves, then the exchange is not providing real investors with any valuable liquidity.

Thus, while micro-quotes help to fulfill the fundamental function of stock markets by encouraging financial agents to provide a lot of bids and offers for a stock, they can hinder the fundamental function of stock markets by reducing the epistemic value of the quotes. This poses a problem of market design as it is necessary to find a balance between the number of quotes and the epistemic value of those quotes.

6.1.4 Market Design Implication: Problem of Balance

The reduced epistemic value of micro-quotes makes it more attractive for agents (such as market makers) to provide orders on those platforms. By attracting a lot of orders at relatively tighter bid/ask spreads, the platforms provide value to real users. However, allowing agents to cancel orders reduces that value. Therefore, a balance must be found. For example, market makers could be allowed to cancel orders only when the market moves against them (which helps them to reduce their risk) in a significant way, that is, the prices of the assets that market makers are holding are moving in a direction that is unfavourable to them.[9] However, they should not be allowed to cancel orders every time the market moved against them (which would be pointless for venues and for end users).

Exchange venues have two possible approaches: to allow for cancellation of quote at any time (1), or not (2).

(1) *Quantity has its own quality*

By attracting a lot of quotes, even if they are cancellable at any time and epistemically devalued (e.g., ghost liquidity), exchanges can make their markets more liquid ('liquidity attract liquidity'). This can lead to tighter ask/bid spreads, which can benefit investors.

(2) *Quality takes (little) time to develop*

Real investors (i.e., non-ghost liquidity) and slower investors might prefer a setting where market makers provide more epistemically reliable quotes, even if this means that the spreads are wider (e.g. a scenario with bid €9.97 and offer € 10.03, instead of a scenario with bid € 9.99 and offer € 10.01). The term 'reliable' here means that market makers would be held to their quotes for a reasonable amount of time, even if the market price moves against them. This would give slower investors more confidence that they would be able to execute their orders at the quoted price.

[9] In this context, I mean that the price movement is sudden and considerable (e.g. the price of an asset to move 10% or more in a single day).

On the flip side, as observed earlier, also the speed bumps, which are the primary solution for micro-quotes, serve as a means to entice HFT traders who place numerous bids and offers while seeking to evade adverse selection.[10] Typically, these HFT traders are market makers providing liquidity even if that liquidity might be 'ghost' to a certain degree. Consequently, the issue of the tension and balance persists in the context of speed bumps as well. On one hand, exchanges strive to acquire an increasing number of "protected" quotes, while on the other hand, they run the risk of diminishing the epistemic value of these quotes by favouring specific traders through intentional design. This means that also speed bumps can affect the markets' efficiency, that is, the capability of markets' of impounding all relevant information into prices.

Thirdly, it is important to note that speed bumps are relevant for non-HFT traders as well (see e.g. Bacidore, 2020). Even if a non-HFT trader is not trying to exploit the short-lived information advantage if micro-quotes, they must still consider speed bumps when submitting large orders. This is because liquidity providers on speed bump venues can learn about large orders by watching fills on other venues during the speed bump's delay period. This can give them time to move their prices and avoid being swept by the large order. To avoid this, non-HFT traders must calibrate and synchronize the arrival of their orders. For example, they could first route orders to speed bump markets and then delay orders to other markets so much that they reach the non-speed bump venues around the time their initial order clears the speed bump. In doing so, liquidity providers will only learn of the other fills after they have completed the original order.

6.2 Conclusion

The choice of the target and time threshold of speed bumps is not a value-neutral or technical decision. It has a significant impact on the structure and agents of financial markets, and it is epistemically and morally loaded: it determines who has access first to information and then is able to profit from market movements, and how efficient and stable financial markets are.

Additionally, we can learn from this how crucial 'reverse finance' and the epistemic and moral values (implicit or explicit) are. Different market designs can be set up for different purposes, which are derived from certain goals and values. This is an exercise in reverse finance, which is theory-laden, meaning that it has specific epistemic and moral values and goals. It is important to make these values and goals explicit, both theoretically and practically, as certain designs favour and

[10] HFTs frequently capitalize on the lack of information among traders by purchasing from them during price declines and selling to them when prices surge. Consequently, this dynamic can result in escalated trading costs for uninformed traders, who are more prone to finding themselves at a disadvantageous end of the transaction (see e.g. Zhang et al., 2008).

attract different actors or orders. For example, if I am a high-frequency trader (HFT) who provides liquidity (market maker), I might trade mostly on certain exchanges and with certain types of orders. On the other hand, if I am an HFT who takes liquidity, I might trade mostly on other exchanges and with other types of orders.

This has important consequences for the structure of the financial markets and their prospects. In fact, in the future we should expect:

(i) A increasing fragmentation of markets in exchanges offering different design, features and advantages.
(ii) a reduction in the epistemic value of quotes due to this fragmentation, as well as in the ability of financial systems to serve their core function even more (see e.g. Chen et al., 2017).

I have shown the epistemic value of a quote in financial markets is tied to the technological innovations and how it has direct implications for market design. The epistemic value of a quote is the agent's belief about the correct price of an asset. This value is time-sensitive, as even milliseconds can make a difference in the accuracy of the quote. In these milliseconds, an agent can access new information about the market and adjust their order accordingly. This gives the agent an advantage over other agents who do not have access to the same information, or who do not react as quickly.

The introduction of micro-quotes has led to a loss of epistemic value in financial markets. Micro-quotes are quotes that can be placed and canceled at any time, and they have a theoretical expiration date measured in milliseconds. This means that the price of an asset can change significantly in the time it takes for a micro-quote to be placed and executed. This makes it difficult for agents to accurately assess the value of an asset, and it gives an advantage to agents who have the fastest technology.

A typical solution to this problem is to introduce speed bumps. Speed bumps are delays that are introduced into the execution of orders. This gives slower agents a chance to react to price changes, and it levels the playing field for all agents. However, the choice of the target and time threshold of speed bumps is not a value-neutral or technical decision. It has a significant impact on the structure and agents of financial markets, and it is epistemically and morally loaded. In fact, we can design different speed bump worlds. I show how these three distinct speed bump worlds can be purposely designed based on predetermined objectives. This process falls under the realm of 'reverse finance' (see Ippoliti, 2020, 2021b), wherein we begin with specific financial market goals and then create a system that can achieve those objectives more effectively.

In conclusion, the epistemic value of a quote in financial markets is time-sensitive, and it has been eroded by the introduction of micro-quotes. One solution to this problem is to introduce speed bumps, but the choice of the target and time threshold of speed bumps is a complex decision with significant epistemic and moral implications, which also affects the stability and the efficiency of the markets that operates on the venues employing those speed bumps.

References

Aoyagi, J. (2020). *The optimal speed to go over speed bumps in financial markets.*

Bacidore, J. (2020). *Algorithmic trading: A practitioner's guide.* TBG Press.

Chami, R., Fullenkamp, C., & Sharma, S. (2010). A framework for financial market development. *Journal of Economic Policy Reform, 13*(2), 107–135.

Chen, H., Foley, S., Goldstein, M., & Ruf, T. (2017). *The value of a millisecond: Harnessing information in fast.* Fragmented Markets. SSRN: https://ssrn.com/abstract=2860359 or https://doi.org/10.2139/ssrn.2860359

De Bruin, B., Herzog, L., O'Neill., Sandberg, J. (2020). Philosophy of money and finance. In Edward N. Zalta (Ed.), *The Stanford Encyclopedia of Philosophy* (Winter 2020 Edition). <https://plato.stanford.edu/archives/win2020/entries/money-finance/>

Fama, E. F. (1970). Efficient capital markets: A review of theory and empirical work. *The Journal of Finance, 25*(2), 383–417.

Fama, E. F. (1991). Efficient capital markets: II. *The Journal of Finance, 46*(5), 1575–1617.

Flanagan, O. J. (1981). Psychology, progress, and the problem of reflexivity: A study in the epistemological foundations of psychology. *Journal of the History of the Behavioral Sciences, 17*, 375–386.

Gabbay, D. M., & Woods, J. (2005). *The reach of abduction.* North-Holland.

Gilbert, M. (2004). Collective epistemology. *Episteme, 1*(2), 95–107.

Gonçalves, J., Kräussl, R., & Levin, V. (2019). *Do 'speed bumps' prevent accidents in financial markets?* CFS Working Paper Series 636. Center for Financial Studies (CFS). https://ideas.repec.org/p/zbw/cfswop/636.html

Harris, L. (2008). *Trading and exchanges: Market microstructure for practitioners.* Oxford University Press.

Ippoliti, E. (2020). *Un filosofo a wall street.* Egea.

Ippoliti, E. (2020a). Mathematics and finance: Some philosophical remarks. *Topoi.* pp. 1–7. https://doi.org/10.1007/s11245-020-09706-1

Ippoliti, E. (2021a). Introduction: Philosophy for finance. *Topoi, 1–7*, 707–713. https://doi.org/10.1007/s11245-021-09759-w

Ippoliti, E. (2021b). Why finance needs philosophy (and vice versa). *Foundations of Science, 27*, 1–18. https://doi.org/10.1007/s10699-021-09804-2

Ippoliti, E. (2023). What is a financial crash? Some remarks on the ontological and epistemic preconditions. *Rivista di Estetica*, Special issue: 'Ontology of finance'. B. Smith, and G. Sansò (Eds.), LXIII:84 (3/2023): pp 7–24. ISSN 2421-5864.

Ippoliti, E. (2024). Do predictions destroy predictability? A study focusing on stock markets. *Logic Journal of IGPL*, 1–19. https://doi.org/10.1093/jigpal/jzae091

Ippoliti, E. (2024a). *Crash point.* Un'analisi filosofica e pratica per capire le crisi finanziarie. Roma, Rubbettino.

Kahneman, D., Slovic, P., & Tversky, A. (Eds.). (1982). *Judgement under uncertainty: Heuristics and biases.* Cambridge University Press.

Khapko, M., & Zoican, M. (2020). How fast should trades settle? *Management Science, 66*(10), 4573–4593.

Levine, M. (2015). *Why do high-frequency traders cancel so many orders?* Bloomberg Opinion.

Levine, M. (2016). *Speed bumps are the hot new thing for exchanges.* Bloomberg Opinion.

Levine, M. (2019). *Buy and sell orders Don't need to be good or evil.* Bloomberg Opinion.

Levine, M. (2019a). *Why exchanges like speed bumps. Bloomberg Opinion.*

Mackintosh, P. (2019). Time is relative: Where trade speed matters, and where it doesn't. *Nasdaq.* https://www.nasdaq.com/articles/time-relative%3A-where-trade-speed-matters-and-where-it-doesnt-2019-05-30

Mackintosh, P. (2020). Time is relativity: What physics has to say about market infrastructure. *Nasdaq,*https://www.nasdaq.com/articles/time-is-relativity:-what-physics-has-to-say-about-market-infrastructure-2020-04-09

Magnani, L. (2009). *Abductive cognition. The epistemological and eco-cognitive dimensions of hypothetical reasoning*. Springer.
Merton, R. C., & Bodie, Z. (1995). A conceptual framework for analyzing the financial environment. In D. Crane, K. A. Froot, S. P. Mason, A. Perold, R. C. Merton, et al. (Eds.), *The global financial system: A functional perspective*. Harvard Business School Press. Ch. 1.
O'Hara, M. (2010). What is a quote? *The Journal of Trading, 5*(2), 10–16.
Popper, K. (1957). *The poverty of historicism*. ARK Paperbacks.
Samuelson, P. (1965). Proof that properly anticipated prices fluctuate randomly. *Industrial Management Review, 6*, 41–49.
Sánchez Serrano, A. (2020). High-frequency trading and systemic risk: A structured review of findings and policies. *Review of Economics, 71*(3), 169–195.
Soros, G. (1987). *The alchemy of finance*. Wiley.
Thicke, M. (2018). Market epistemology. *Synthese, 195*(12), 5571–5594.
Williams, E. E., & Dobelman, J. A. (2020). *A random walk to nowhere: How the professors caused a real "fraud-on-the-market"*. World Scientific Publishing.
Zhang, M. Y., Russell, J. R., & Tsay, R. S. (2008). Determinants of bid and ask quotes and implications for the cost of trading. *Journal of Empirical Finance, 15*(4), 656–678.

Chapter 7
Why We Need a Computational Ontology of Finance (and How Philosophers Can Help Build It)

Gloria Sansò

Abstract The financial crisis of 2007–2008 showed, among other things, that the information technology in the financial sector was inadequate to assist institutions in producing reliable reports and keeping track of the securities issued. This, in turn, compromised the chances of correctly estimating financial risk. In this paper, I explain how a well-designed computational ontology of finance could help improve data management. I also argue that this ontology should be realism-based and conformant to Basic Formal Ontology (BFO). This choice, however, does not come without complications because it requires some preliminary reflections on the nature of the entities involved. Fortunately, some philosophical literature can be a valuable starting point for these reflections; in the paper, this is shown by considering money as a case study.

7.1 Introduction

One of the most astonishing facts about the financial crisis of 2007–2008 is that, while some people had a clue that the crisis was coming (Beattie, 2008), most economists did not realize, in time, what was going on. Why did that happen? There are different possible answers. One may argue, for example, that social sciences, and more specifically some branches of economics, are based on some sort of metaphysical misunderstanding about the building blocks of the social world, and this prevents scholars from making proper predictions and understanding when things are turning for the worse (Epstein, 2015, pp. 2–10). Others argue that the problem is more methodological than metaphysical: by working almost exclusively on models, economists neglect what is going on in real life (Sylos Labini & Caprara, 2017, p. 68).

G. Sansò (✉)
University at Buffalo (SUNY), Buffalo, NY, USA
e-mail: glrsanso@gmail.com

E. Ippoliti et al. (eds.), *Philosophy and Finance: Ten Open Questions*, Synthese Library 504, https://doi.org/10.1007/978-3-031-96510-4_7

Although there may be some truth in both these views, it must be said that, in general, a proper estimation of the systemic risk is challenging for another reason as well. The systemic risk is the potential collapse of the entire financial system (Bodie et al., 2018, p. 21), which entails the "disruption of the market's ability to facilitate the flow of capital that results in the reduction in the growth of the global GDP" (Fouque & Langsam, 2013, p. xxi). Its precise assessment is possible *only if* one has a reliable picture of the system as a whole, but this big picture requires a plethora of data that, in the years prior to the crisis, no one could have. On the one hand, many transactions involved shadow banks, namely financial intermediaries, that were outside the scope of government regulations and, as such, characterized by little transparency (Schoen, 2017, p. 811). As a consequence, "the CDOs market had turned into a vast, *opaque* spider web of deals in which banks, shadow banks, and brokers alike had become dangerously ensnared, interlinked by fear" [emphasis added] (Tett, 2009, p. 226).

On the other hand, there was a more general problem afflicting all the financial institutions, including the largest banks, that is:

> banks' information technology (IT) and data architectures were inadequate to support the broad management of financial risks. Many banks lacked the ability to aggregate risk exposures and concentrations quickly and accurately at the bank group level, across business lines, and between legal entities. Some banks were unable to manage their risks properly because of the weak risk data aggregation capabilities and risk reporting practices (Basel Committee, 2013, p. 1).

In the financial sector, thus, informatic systems failed in two different ways. First of all, they were inadequate to assist each bank in producing reliable reports by collecting data about itself. Secondly, the informatic systems of each institution were incapable of working with those of the others and properly exchanging information; this drastically undermined the possibility of keeping track of the securities issued. When, for example, Lehman Brothers collapsed, no one knew how many CDOs were actually in the market, which of those were toxic, and who held them (Helleiner, 2011, p. 71; Dombret, 2013, p. 2; Chadha, 2016, p. 3).

There are several computational approaches that can be used to improve the informatic systems in the financial sector (Hughes et al., 2013); among them, semantic technologies seem to be particularly promising (Butler et al., 2015; Chen, 2015). In semantic technologies, ontology is the key component; therefore, creating a successful ontology of finance is of paramount importance in order to increase transparency and facilitate regulatory compliance by financial institutions.

In this paper, after introducing semantic technologies and the role of the ontology (Sect. 2), I present the benefits (and the challenges) of a realism-based ontology of finance (Sect. 3), and I show how philosophy can help in building it by considering money as a case study (Sect. 4).

7.2 Semantic Technologies

The expression "semantic technologies" denotes those methods and tools that make the content of the information accessible to machine processes. In doing so, they greatly enhance the capabilities of the information systems. If machines are able to interpret the content of a piece of information, then they can make inferences and derive knowledge that, although not in the system, can be inferred from the information in the system (Fürber, 2016, pp. 65–66). Suppose, for example, that a system has the information that a certain firm *A* owns firm *B*, and that *B* has a controlling interest in firm *C*, where having a controlling interest in *C* implies owning more than half of *C*'s voting shares. If this system has access to the content of "owning" and "having a controlling interest", then it can infer that *A* has a controlling interest in *C*. Secondly, if machines are able to interpret the content of a piece of information, then a system can use information present in another system (and *vice versa*), even when these systems may have different data structures, formats, and even vocabularies (SICoP, 2005, pp. 37–38). Semantic technologies can thus promote interoperability, that is, the ability of different systems to exchange information, search, query, and reason across multiple data sources. In order to better understand how all this works, it may be helpful to consider more details about the elements involved.

As shown in Fig. 7.1, in each business unit, there are data, metadata, and an ontology. Data can be both structured and unstructured. Structured data are

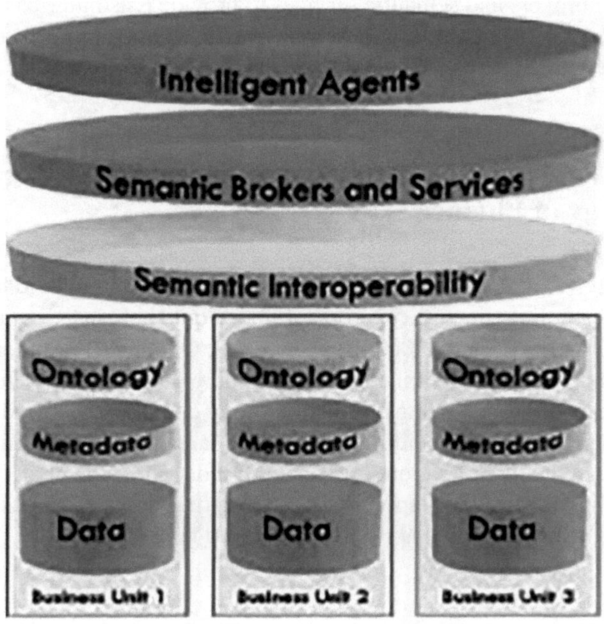

Fig. 7.1 Semantic Technologies (Cregan, 2008, p. 59)

quantitative and are stored in tables, such as an SQL database or an Excel file; some examples of structured data include dates, names, and prices. Unstructured data, instead, do not have a predefined format and are embedded in things like audio, video, or a Word file (Weglarz, 2004).

Metadata is data about data, meaning that it describes the relevant information about data. So, for example, the content of this essay is the data, whereas the size of the file, the author, and the time it was last edited are metadata. Metadata serves as a bridge between data and ontology: by organizing metadata, an ontology helps identify and retrieve data.

In information and computer science, the term "ontology" denotes a formal representation of a portion of reality; such a representation includes a collection of terms: some of them refer to classes, whereas others refer to the relations between them. Among these relations, the *is_a* relation is particularly relevant in that it serves as a fundamental reference point for creating a taxonomy, namely a hierarchy consisting of classes and sub-classes, which is the backbone of any ontology (Arp et al., 2015, p. 1; Guarino et al., 2009, p. 2). Importantly, an ontology provides, for each of its terms, an accurate, unambiguous, and consistent semantics (Gruber, 1995, p. 11; Guarino, 1998, pp. 6–7) that is expressed in both human and machine-readable formats. Because of these characteristics, an ontology is said to have three main utilities: (i) to facilitate the communication between human beings; (ii) to promote interoperability between computer systems, and (iii) to improve the quality of pieces of software (Jasper & Uschold, 1999, p. 17).

Ontologies, thus, are the crucial element in semantic technologies because they make semantic interoperability possible; semantic interoperability, in turn, allows semantic brokers and semantic services to provide reasoning-based services; artificial agents, then, can compose these services to perform more complex tasks, such as analyzing documents and making decisions (see, for example, Joshi et al., 2019; Turner et al., 2019).

7.3 Ontology of Finance

Based on the above, it seems evident that a well-designed computational ontology of finance, by promoting semantic interoperability, would help in assessing systemic risk (see, for example, Ye et al., 2009; Bennett, 2013; Organ, 2018) and, in so doing, it could solve the problems mentioned at the beginning of this chapter. However, this is not all. Indeed, different portions of this ontology could also be used for several more specific tasks, such as providing support in financial planning (Bunnell et al., 2020), assisting in the identification of financial frauds (Kingston et al., 2004; Zhao et al., 2004; Chmielewski & Stąpor, 2018), and rating companies based on their financial quality (Shue et al., 2009; Martin et al., 2011).

The question, now, is: how can one build such a computational ontology? Different methodologies can be employed: the attempts made so far to create an ontology of finance[1] are concept-based. There are, however, some general concerns about the conceptualist approach; for that reason, in what follows, I argue that a reality-based ontology of finance would be an exciting alternative and that philosophy can help overcome some of the challenges involved in this endeavor.

7.3.1 Concept-Based Ontologies

In computer and information science, the conceptualist approach is grounded on the idea that, in order to build a valuable ontology, it is not necessary to commit to a definite metaphysic of the world. The categories employed in an ontology are simply "cognitive artifacts ultimately dependent on human perception, cultural imprints and social conventions" (Gangemi et al., 2002, p. 167). Consequently, an ontology does not represent things in reality; it just represents our concepts of these things.

Unfortunately, the conceptualist approach is problematic for several reasons.[2] One of them is that such an approach deprives us of a standard to evaluate ontologies. Concepts are indeed subjective, and if an ontology merely represents concepts, it is unclear on what basis to evaluate them. Of course, an ontology can be evaluated based on the coherence of its concepts or, alternatively, on how common these concepts are among people. However, this evaluation would still have a certain degree of subjectivity. Things entirely change if one assumes that an ontology represents things in reality because, if this is the case, then one can assess an ontology based on how accurately it represents what there is out there (Smith, 2004, p. 76).

One may counterargue that it is possible to evaluate ontologies based on their utility, meaning based on their ability to perform the task they are built for. Therefore, what ontologies represent (concepts or things in reality) and, more generally, the underlying metaphysical arguments, are secondary factors when assessing them. Ontologies are good when they successfully do what they are supposed to do and bad when they do not (Guarino, 2017, p. 11).

While an ontology certainly has to be able to perform the task it is built for, there are at least two reasons why this criterion, alone, is insufficient to evaluate ontologies properly. First, in information and computer science, as shown earlier, the aim is not simply to have ontologies performing specific tasks but to have systems able to communicate with each other. Interoperability can be achieved more easily if ontologies are consistent, meaning if they are created by referring to the

[1] I am referring to the Financial Industry Business Ontology (FIBO), a well-known and comprehensive project, and to the Ontology Network in Finance and Economics (OntoFINE).

[2] For a more comprehensive overview of the main criticisms of the conceptualist approach, see Smith (2004) and Arp et al. (2015, pp. 5–15).

same benchmark, which is reality (see, for example, Smith et al., 2007; Arp et al., 2015, p. 48; Kulvatunyou et al., 2018). Second, it is only sometimes possible to know in advance for which task an ontology will be used in the future, and this is especially true for reference ontologies. Indeed, the purpose of a reference ontology is to represent all there is in a specific domain under the assumption that, at some point, someone will use part of it to build an application ontology, an ontology with a specific task. The utility of a reference ontology lies in its ability to stand the test of time and to be reused by as many application ontologies as possible. Interestingly, it has been proven that realism-based ontologies better serve this goal than those representing concepts (Merrell et al., 2021, p. 3).

7.3.2 Building a BFO-Conformant Ontology of Finance

Basic Formal Ontology (BFO) is the upper-level ontology[3] developed by Barry Smith and associates; its last version, released in 2020, fulfills the requirements for top-level ontologies in ISO/IEC 21838-1, and it is thus generally acknowledged as being able to support interoperability among heterogeneous information systems.[4] Since philosophical theories were (and still are) of great support for developing BFO and its extensions, BFO can be considered the first product of philosophy to become an industry standard (Jansen & Brochhausen, 2022).

In BFO, entities are divided into two main categories: *continuants* and *occurrents*. *Continuants* are entities that continue to exist through time (some examples include my desk, the content of a book, the function of a clock); *occurrents* are entities that occur or happen (some examples include a ceremony, a war, a storm) (Arp et al., 2015, p. 87). Continuant and occurrent entities are further categorized (see Fig. 7.2).

Currently, BFO is used as a foundation for 420 ontologies, and it is involved in many projects[5] belonging to different fields, such as biomedicine, national security, and industry. BFO and BFO-conformant ontologies share the same methodology called ontological realism. According to this view, a good ontology is *not* a representation of concepts but a representation of reality as our best current science describes it (Smith, 2004, p. 76; Arp et al., 2015, p. 48). While it is true that, sometimes, scientific statements are false, meaning that they do not describe reality as it really is, scientific statements are typically accurate, and they are thus the best candidates for being the benchmark to create and evaluate ontologies.

---

[3] An upper-level ontology is an ontology providing very general terms, such as "object", "event", and "property", which are common to any domain. In contrast, domain ontologies are ontologies representing specific portions of reality; they thus provide terms that are peculiar to those areas. A domain ontology is typically an extension of an upper-level ontology.

[4] https://www.iso.org/standard/74572.html (last accessed May 17, 2023).

[5] http://basic-formal-ontology.org/users.html (last accessed: May 17, 2023).

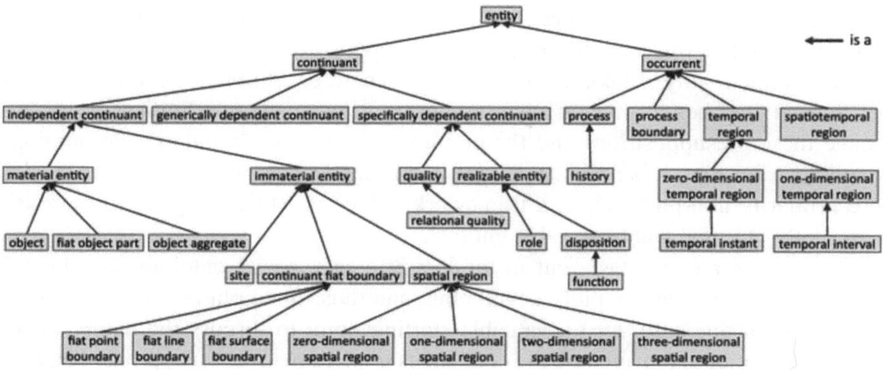

Fig. 7.2 BFO 2020: is_a hierarchy

Each term in a realist ontology refers to a universal or an individual. "Universals are entities in reality that are responsible for the structure, order, and regularity - the similarities - that are to be found there. To talk of universals is to talk of what all members of a natural class or natural kind such as *cell*, or *organism*, or *lipid*, or *heart* have in common" (Arp et al., 2015, p. 13). So, for example, the term "heart" refers to the collection of qualities that an object must have in order to be a heart, while a specific heart, such as Mary's heart, is an individual instantiating *that* universal. It is important to point out that a universal exists only insofar as at least one individual is instantiating it. This means that, although one can perhaps list the qualities that a thing must have in order to be a unicorn, "unicorn" cannot be a term in a realist ontology because no existing individual instantiates that universal.

The above quotation refers to natural kinds and natural entities in general. However, ontological realism can also be used as a methodology to represent social entities (see Arp et al., 2015, p. 44). Thus, the BFO framework lends itself to being a useful tool to build an ontology of finance and avoid the issues of the conceptualist approach. On the other hand, this choice does not come free of complications. Creating realism-based ontologies is *per se* more complex than building concept-based ontologies: because of their ontological commitment to entities in reality, realist ontologies cannot include, for example, terms such as "prevented abortion", "absence of metastases", or "surgical biopsy not taken", since they do not, strictly speaking, denote existing things. However, in some contexts, it may be necessary to represent these states of affairs in order to keep track of the related data; in order to do that, one has to come up with a solution and, for example, introduce new relations, such as the *lack* relation (for more details about this, see Ceusters et al., 2007; Schuler & Ceusters, 2018).

In addition to this general difficulty, there may be field-specific challenges, such as those that one must face in representing the social world. While it may be enough to leaf through a science textbook in order to understand what there is in the natural world and how to represent its phenomena, the same does not apply to the social world. Although an economics textbook explains many phenomena, such an explanation is often based on implicit presuppositions about the bricks forming the economic reality (Mäki, 2001, pp. 3–14). Therefore, to build a realist ontology of finance (and benefit from it), it is necessary to take some preliminary steps and make these presuppositions and the entities one needs to include in the ontology explicit. After that, it is necessary to explain how to represent these entities, which are primarily non-physical, in a framework, which is BFO, designed to represent especially physical and observable entities.

This is not an easy task, but in the last 30 years, social ontology as a branch of philosophy has been rapidly developing, and thus, some philosophical literature can offer (in this case, too) a valuable starting point to discuss the nature of the financial entities and overcome the challenges implied by a realist approach. In what follows, I shall show this by considering money as a case study: although providing a complete theory goes beyond the scope of this chapter, I outline what can be seen as a theoretical foundation for the representation of money in a BFO-compatible ontology of finance.

7.4 Money

The nature of money has been investigated over the centuries by prominent scholars of various disciplines, such as philosophy, anthropology, economics, and sociology. In the last few decades, money has also been one of the most debated topics in social ontology; in what follows, I discuss some of the views that emerged.

7.4.1 Money and Its Representations

John Searle is well-known for introducing the formula *X counts as Y in C*, through which he aims to account for institutional facts (or, at least, for some of them) (1995, 2010). A certain object is a dollar bill because, he says, people collectively assign a new status to that object, that is the status of dollar bill, and, in virtue of this new status, this object can perform certain functions in a given context, such as the function of means of exchange in the United States. One may thus say that the piece of paper in my pocket *counts as* a dollar bill in the United States.

Although this formula has some plausible appeal, it has been challenged in many ways; one of the most compelling objections is advanced by Barry Smith. Smith (2003, pp. 24–27) points out that there are many social entities that do not coincide with any physical object X; some examples include debts, obligations, rights, and digital money. "The blips in the bank's computer merely *represent* money, just as the deeds to your property merely *record* or *register* the existence of your property right. The deed is not identical with your property right and nor does it count as your property right" (Smith, 2003, p. 20). These entities are called by Smith “free-standing Y terms” or “quasi-abstract” because, like abstract entities, they are non-physical and non-psychological but, on the other hand, they are tied to time, so they are historical (Smith, 2008, p. 37).

Based on that, it seems that, in addition to things that are *actually* money, such as coins and banknotes, there are things, such as the blips in the bank’s computer or the ink on a bank statement, that just *represent* money. If one assumes that “representing” means being about some portion of reality (Arp et al., 2015, p. 3) and that coins and banknotes are money, then “representing money” means being about some banknotes and coins (maybe those banknotes and coins that I have the right to withdraw from the bank based on my most updated bank statement). The problem with this view is that, as it is, it cannot accommodate the case of Central Bank Digital Currencies (CBDCs), which could, over the long term, completely supplant paper cash.[6] Although governments do not have concrete plans in that direction, one can suppose that, at some point, banknotes and coins will become obsolete, and transactions will occur electronically only. If so, then it is not clear what a bank statement would represent since banknotes and coins would cease to exist as money. A possible way to solve this problem is to say that money is something other than the objects used as a means of exchange. Therefore, a bank statement does not represent banknotes and coins, but something else. It remains to be clarified, though, what “something else” is.

7.4.2 *Institutions and Causal Powers*

Guala (2020) argues that institutions are ontologically more fundamental than the devices used within them; this means, for example, that money-as-an-institution has ontological priority over money-as-an-object, that is the object used as money, such as coins and banknotes. Consequently, to understand what money is (and thus, what a bank statement represents), it is a good idea to consider money-as-an-institution first. Institutions are defined as “*rules in equilibrium*, or rules that people are motivated to follow” (Guala, 2020, p. 272). Indeed, while there is no doubt that rules form an institution and that some sort of authority explicitly issues some of them, these rules alone do not make people observant. In order to follow the rules, people need an incentive, as the fact that following the rules makes their lives easier. A person living in the US, for example, is motivated to use dollar bills not (just) because there are some laws prescribing this but because, since most people in the US are using dollar bills, this person finds it convenient to use dollar bills as well. In this context, using dollar bills allows this person to enjoy a number of benefits, such as participating in smooth transactions, sharing a unit of account, and so on.

[6] The same applies to cryptocurrencies, such as Bitcoin (assuming that one wants to consider them as money).

"If the system of expectations is coherent, the institution is self-sustaining, and our lives are a lot simpler if we follow the rules" (Guala, 2020, p. 273).

If that is the case, then there is little point in investigating the nature of banknotes and coins to understand what money is because (i) they are just two monetary devices among many others, and (ii) monetary devices become ontologically interesting once they are placed within an institution since that would help to understand what they do. Instead of merely focusing on banknotes and coins, it is much more promising to zoom out and better understand the relations between these objects and the institution to which they belong. In a way, this is what Mäki tries to do: rather than specific instantiations, he focuses on "moneyhood", the collection of properties shared by *all* monetary devices, and defines it as "a set of *institutionally* and *practically* sustained causal powers" [emphasis added] (Mäki, 2021, p. 253).

In metaphysics, causal powers are dispositional properties, meaning properties that, under certain circumstances, make objects behave in a certain way; some examples of dispositional properties include being inflammable, being fragile, and being soluble. Causal powers are typically used to offer an account of causation: the solubility of a sugar cube, that is, the disposition of a sugar cube to dissolve itself, accounts for the change involving the cube when it is placed in non-saturated water. Mäki believes that something similar can be claimed of the objects instantiating moneyhood, and this is also reflected by the fact that money-as-an-object is commonly said, for example, to have purchasing *power*. Unlike the powers a sugar cube possesses, the powers exhibited by monetary devices do not derive, of course, from their physical structure but rather from their placement within an institution, which attributes those powers to these objects. In this framework, a bank statement does not represent banknotes and coins but the collection of powers somehow carried by these monetary devices.

This view has, I think, two main benefits. First, it recognizes the role of money-as-an-institution in determining the properties possessed by money-as-an-object. Second, it accounts for the modal dimension of monetary devices: like sugar cubes, these devices have potentialities, such as that of facilitating people's exchanges, that manifest themselves only when certain conditions occur. On the other hand, Mäki's view seems to imply the existence of a "thing" carrying these powers, and this is highly problematic in the case of digital currencies, such as CBDCs, where there is no money-as-an-object. In addition (and more generally), appealing to causal powers in order to account for social phenomena is a controversial choice that brings with it several legitimate concerns.[7] The most relevant, for the purpose of this chapter, is that causal powers (and dispositional properties in general) are typically said to be internal,[8] meaning that they are possessed by an object just in virtue of itself. This is clearly incompatible with the case of money since the objects instantiating moneyhood possess their qualities just in virtue of an institution and not

[7] These concerns are discussed by Wahlberg (2018).

[8] See, for example, Harré (1970); Molnar (2003); Bird (2007); Marmodoro (2017); Williams (2019).

in virtue of themselves; these qualities are thus external. While this does not seem to be the biggest problem of all (Mäki himself, as well as a minority of philosophers, does not believe that causal powers *have* to be internal properties), it prevents one from using Mäki's theory as a theoretical foundation to represent money in the BFO framework. Indeed, in BFO, like in most of the philosophical literature, an entity is a disposition if it is internally grounded, meaning grounded in the physical make-up of an object (Arp & Smith, 2011, pp. 6–7) and, clearly, the powers of money do not fulfill this condition.

7.4.3 Information and Roles

My proposal is to give an account of money by considering *deontic* powers, meaning abilities that an individual has (or does not have) in virtue of her social role and some regulations. According to Searle (1995, pp. 84–85), some of these deontic powers are positive, whereas others are negative. Rights and permissions are instances of the former; they are *positive* because they make new courses of action possible. Obligations and prohibitions are, instead, instances of the latter; they are *negative* because they limit people's actions.

Deontic powers allow us to account for modality in the social world without, however, the complications implied by causal powers. Consider, for example, my right to vote: in virtue of this deontic power, when certain conditions are obtained, meaning that there are elections in my home country, I have the possibility to perform a certain action, that is, voting. Something similar can be said about money. In virtue of the right to use a certain currency, when certain conditions are obtained, meaning that there is a commercial opportunity and I have a certain amount of that currency, I have the possibility to perform a certain action, that is, purchasing. Like Mäki, thus, I claim that money involves powers that are "institutionally and practically sustained"; unlike Mäki, I claim that these powers are (i) *deontic,* not *causal*, and (ii) possessed by individuals, not by monetary devices.

The nature of these deontic powers remains to be clarified. Based on our ordinary talk, it seems that expressions such as "right to vote" or "property right" denote (somehow abstract) things that people can actually own. Following Donohue (2020, pp. 139–158), however, I opt for a different approach, which is, I believe, ontologically more reasonable insofar as it refrains from asserting the existence of metaphysically controversial entities. I argue that deontic powers are not individual things but states of affairs, which include regulations, people holding certain roles, and the relations between these two things. The sentence "I have the right to vote in Italy", for example, does not refer to a situation where there is a thing, the right to vote, owned by me. Rather, this sentence refers to a situation where (i) I fulfill the conditions to have the role of voter, (ii) there is some regulation, such as the Constitution, that defines the actions I am allowed to perform in virtue of this role, and (iii) people around me are motivated to follow that regulation and allow me to perform the actions provided for.

The same can be applied to money. The sentence "I have the right to use currency X" does not refer to a situation where there is an individual thing that can be actually owned by people. Such an expression refers, instead, to a state of affairs that includes the rules forming the institution of money as well as those governing the commercial exchange, people holding certain roles, and the relations between these things. Therefore, the sentence "I have the right to use currency X", refers to a situation where (i) I fulfill the conditions to have the role of owner of currency X, (ii) there is some regulation that allows me to perform some actions, such as purchasing, in virtue of my role and by means of currency X, and (iii) people around me are motivated to follow that regulation and allow me to perform the actions provided for.

What I have said so far about deontic powers accounts for money-as-an-institution, but what about money-as-an-object? I argue that monetary devices, such as coins and banknotes, can be seen as carriers of information: they specify an amount of currency and, together with other information, such as the purchasing power of that specific currency, define the range of things that one can acquire by means of them. In the case of a CBDC, where there are no banknotes and coins, one may assume the existence of documents such as bank statements that, similarly to banknotes and coins, convey information about the amount of currency possessed by each individual. Although these documents are not instances of money-as-an-object, and they cannot be exchanged as banknotes and coins, they represent the possibilities implied by the amount of currency the account holder possesses. This, along with the rules forming the institution (and the willingness of people to follow them), is enough to put into existence a state of affairs where people have (or do not have) the ability to participate in transactions, even in the absence of banknotes and coins.

My proposal, which prioritizes money-as-an-institution and characterizes money-as-an-object in terms of carrier of information, effectively addresses the challenges related to the absence of money-as-an-object illustrated in 4.1 and 4.2. It also clarifies what documents such as bank statements represent. The second advantage of my proposal is that it relies on types of entities that can be easily included in a BFO-conformant ontology. More specifically, my account relies mostly on pieces of information carried by physical objects and roles held by people. In the context of BFO, a piece of information can be seen as an abstract pattern, such as a pattern of letters. This abstract pattern is concretized in some of the qualities possessed by the material support that is carrying the information; some examples include ink marks, in the case of a piece of paper, and magnetic traces, in the case of the disk of a computer (Arp et al., 2015, pp. 105–107). Also, a piece of information is always about something (Ceusters & Smith, 2015, p. 2). In the case of regulations, bank statements, and monetary devices, the information they convey is, for example, about the actions some people can take and/or the amount of currency they possess. In BFO, role is (unlike disposition) an *externally-grounded* entity, meaning an entity that something has because of an external circumstance (Arp et al., 2015, pp. 99–101). This means, for example, that an individual has the role of owner because she meets some institutional requirements and, in virtue of this, she has the possibility to perform actions that would be otherwise closed to her, such as participating in a transaction.

7.5 Conclusions

In this paper, I have shown that, in the financial sector, a more efficient information technology would facilitate data management and assist institutions in producing more reliable reports, keeping track of the securities issued, and better assessing systemic risk. One way to make this improvement is by building a computational ontology of finance. I have argued that since realism-based ontologies have often proven more successful than concept-based ontologies, building a realism-based ontology of finance would be an interesting endeavor. In particular, I have suggested Basic Formal Ontology (BFO) as the appropriate framework for building such an ontology. This choice, however, carries with it some challenges insofar as it requires some preliminary steps in order to identify a way to account for financial entities that is satisfactory as well as BFO-conformant. By considering money as a case study, I have shown that some philosophical literature is a valuable support in that process. Based on some recent work on the ontology of money, I have accounted for money and its modal dimension in terms of deontic powers and deontic powers in terms of states of affairs populated by two main entities: the roles possessed by the individuals and the pieces of information carried by regulations and monetary devices.

References

Arp, R., & Smith, B. (2011). Function, role and disposition in Basic Formal Ontology. *Nature Precedings*, 1–4.

Arp, R., Smith, B., & Spear, A. (2015). *Building ontologies with basic formal ontology*. MIT Press.

Basel Committee. (2013). *Principles for effective risk data aggregation and risk reporting*. https://www.bis.org/publ/bcbs239.pdf

Beattie, A. (2008, November 14). *Good question, Ma'am*. Financial Times.

Bennett, M. (2013). The financial industry business ontology: Best practice for big data. *Journal of Banking Regulation, 14*(3–4), 255–268.

Bird, A. (2007). *Nature's metaphysics: Laws and properties*. Oxford University Press.

Bodie, Z., Kane, A., & Marcus, A. J. (2018). *Investments* (11th ed.). McGraw-Hill Education.

Bunnell, L., Osei-Bryson, K., & Yoon, V. (2020). Development of a consumer financial goals ontology for use with FinTech applications for improving financial capability. *Expert Systems with Applications, 165*(3), 1–14.

Butler, T., Abi-Lahoud, E., & Espinoza, A. (2015). Designing semantic technologies for regulatory change management in the financial industry. *ICIS 2015 Proceedings, 8*, 1–11.

Ceusters, W., Elkin, P., & Smith, B. (2007). Negative findings in electronic health records and biomedical ontologies: A realist approach. *International Journal of Medical Informatics, 76*, 326–333.

Ceusters, W., & Smith, B. (2015). Aboutness: Towards foundations for the information artifact ontology. *Proceedings of the Sixth International Conference on Biomedical Ontology (ICBO) CEUR, 1515*, 1–5.

Chadha, P. (2016). What caused the failure of Lehman brothers? Could it have been prevented? How? Recommendations for going forward. *International Journal of Accounting Research, s1*, 1–5.

Chen, J. (2015). *Evaluation of application of ontology and semantic technology for improving data transparency and regulatory compliance in the global financial industry* (Master's thesis). Massachusetts Institute of Technology.
Chmielewski, M., & Stąpor, P. (2018). Hidden information retrieval and evaluation method and tools utilising ontology reasoning applied for financial fraud analysis. *MATEC Web of Conferences*, 210.
Cregan, A. (2008). *Weaving the semantic web: Contributions and insights* (Doctoral dissertation). University of New South Wales Sydney].
Dombret, A. (2013). Lehman lessons. *International Economy, Fall 2013*, 34–36.
Donohue, B. (2020). *Social ontology and social normativity* (Doctoral Dissertation). University at Buffalo.
Epstein, B. (2015). *The ant trap: Rebuilding the foundations of the social sciences*. Oxford University Press.
Fouque, J., & Langsam, J. A. (2013). Introduction. In J. Fouque & J. A. Langsam (Eds.), *Handbook on systemic risk* (pp. xx–xxv). Cambridge University Press.
Fürber, C. (2016). *Data quality management with semantic technologies*. Springer.
Gangemi, A., Guarino, N., Masolo, C., Oltramari, A., & Schneider, L. (2002). Sweetening ontologies with DOLCE. In A. Gómez-Pérez & V. R. Benjamins (Eds.), *Knowledge engineering and knowledge management: Ontologies and the semantic web. EKAW 2002 (Lecture Notes in Computer Science)* (pp. 166–181). Springer.
Gruber, T. R. (1995). A translation approach to portable ontologies. *Knowledge Acquisition, 5*(2), 199–220.
Guala, F. (2020). Money as an institution and money as an object. *Journal of Social Ontology, 6*(2), 265–279.
Guarino, N. (1998). Formal ontology and information systems. In N. Guarino (Ed.), *Formal ontology in information systems* (pp. 3–15). IOS Press.
Guarino, N. (2017). BFO and DOLCE: So far, so close. *Cosmos+Taxis, 4*(4), 10–18.
Guarino, N., Oberle, D., & Staab, S. (2009). What is an ontology? In S. Staab & R. Studer (Eds.), *Handbook on ontologies* (pp. 1–17). Springer.
Harré, R. (1970). Powers. *The British Journal for the Philosophy of Science, 21*(1), 81–101.
Helleiner, E. (2011). Understanding the 2007–2008 global financial crisis: Lessons for scholars of international political economy. *Annual Review of Political Science, 14*, 67–87.
Hughes, E., Rosenthal, A., & Worrell, C. (2013). Enabling data analysis for addressing systemic risk. In J. Fouque & J. A. Langsam (Eds.), *Handbook on systemic risk* (pp. 852–868). Cambridge University Press.
Jansen, L., & Brochhausen, M. (2022, September 28). *Begriffe in Schichten: Die 'Basic Formal Ontology' erbringt Dienstleistungen für die Medizin*. Das wurde jetzt mit der Anerkennung als Industriestandard gewürdigt. Frankfurter Allgemeine Zeitung.
Jasper, R., & Uschold, M. (1999). A framework for understanding and classifying ontology applications. In *Proceedings 12th international workshop on knowledge acquisition, modelling, and management* (pp. 16–21).
Joshi, K., Pande, K., & Mittal, S. (2019). A semantic approach for automatic knowledge in policies of cyber insurance services. In 2019 *IEEE international conferences on web services* (pp. 33–40).
Kingston, J., Schafer, B., & Vandenberghe, W. (2004). Towards a financial fraud ontology: A legal modelling approach. *Artificial Intelligence and Law, 12*(4), 419–446.
Kulvatunyou, B., Wallace, E., Kiritsis, D., Smith, B., & Will, C. (2018). The industrial ontologies foundry proof-of-concept project. *APMS 2018: Advances in Production Management Systems.* Smart Manufacturing for Industry *4.0,* pp. 402–409.
Mäki, U. (2001). Economic ontology: What? Why? How? In U. Mäki (Ed.), *The economic world view: Studies in the ontology of economics* (pp. 3–14). Cambridge University Press.
Mäki, U. (2021). Reflections on the ontology of money. *Journal of Social Ontology, 6*(2), 245–263.
Marmodoro, A. (2017). Aristotelian powers at work: Reciprocity without symmetry in causation. In J. D. Jacobs (Ed.), *Causal powers* (pp. 57–76). Oxford University Press.

Martin, A., Manjula, M., & Venkatesan, P. (2011). A business intelligence model to predict bankruptcy using financial domain ontology with association rule mining algorithm. *International Journal of Computer Science Issues, 8*(3), 211–218.
Merrell, E., Kelly, R., Kasmier, D., Smith, B., Brittain, M., Ankner, R., Maki, E., Heisey, C. W., & Bush, K. (2021). Benefits of realist ontologies to systems engineering. In *8th international workshop on ontologies and conceptual modelling (OntoCom)* (pp. 1–10).
Molnar, G. (2003). *Powers: A study in metaphysics*. Oxford University Press.
Organ, J. (2018). The control of human factors in catastrophic financial systemic risk: A case study of the Irish banking crisis. *IFAC-PapersOnLine, 51*(30), 580–585.
Schoen, E. J. (2017). The 2007–2009 financial crisis: An erosion of ethics: A case study. *Journal of Business Ethics, 146*, 805–830.
Schuler, J. C., & Ceusters, W. M. (2018). The problems of realism-based ontology of design: A case study in creating definitions for an application ontology for diabetes camps. *AMIA Annual Symposium Proceedings, AMIA Symposium, 2017*, 1517–1526.
Searle, J. (1995). *The construction of the social reality*. Free Press.
Searle, J. (2010). *Making the social world*. Oxford University Press.
Shue, L., Chen, C., & Shiue, W. (2009). The development of an ontology-based expert system for corporate financial rating. *Expert Systems with Applications, 36*(2), 2130–2142.
SICoP. (2005). *Introducing Semantic Technologies and the Semantic Web Vision.*https://staff.icar.cnr.it/pontieri/didattica/SIIR/lezioni/SemanticWeb.pdf
Smith, B. (2003). John Searle: From speech acts to social reality. In B. Smith (Ed.), *John Searle* (pp. 1–33). Cambridge University Press.
Smith, B. (2004). Beyond Concepts: Ontology as Reality Representation. In A. Varzi, L. Vieu (eds.), *Proceedings of Fois 2004: International Conference on Formal Ontology and Information systems* (pp. 1–12).
Smith, B. (2008). Searle and de Soto: The new ontology of the social world. In B. Smith, D. Mark, & I. Ehrlich (Eds.), *The mystery of capital and the construction of social reality* (pp. 35–51). Open Court.
Smith, B., Ashburner, M., Rosse, C., Bard, J., Bug, W., Ceusters, W., Goldberg, L. J., Eilbeck, K., Ireland, A., Mungall, C. J., OBI Consortium, Leontis, N., Rocca-Serra, P., Ruttenberg, A., Sansone, S. A., Scheuermann, R. H., Shah, N., Whetzel, P. L., & Lewis, S. (2007). The OBO foundry: Coordinated evolution of ontologies to support biomedical data integration. *Nature Biotechnology, 25*(11), 1251–1255.
Sylos Labini, F., & Caprara, S. (2017). On the description of financial markets: A physicist's viewpoint. In E. Ippoliti & P. Chen (Eds.), *Methods and finance: A unifying view on finance, mathematics and philosophy* (pp. 67–71). Springer.
Tett, G. (2009). *Fool's gold: How unrestrained greed corrupted a dream, shattered global markets and unleashed a catastrophe*. Little, Brown and Company.
Turner, C. J., Emmanouilidis, C., Tomiyama, T., Tiwari, A., & Roy, R. (2019). Intelligent decision support for maintenance: An overview and future trends. *International Journal of Computer Integrated Manufacturing, 32*(10), 936–959.
Wahlberg, T. H. (2018). Causal powers and social ontology. *Synthese, 197*(3), 1357–1377.
Weglarz, G. (2004). Two worlds of data - unstructured and structured. *DM Review*.
Williams, N. (2019). *The powers metaphysic*. Oxford University Press.
Ye, K., Wang, S., Yan, J., Wang, H., & Miao, B. (2009). Ontologies for crisis contagion management in financial institutions. *Journal of Information Sciences, 35*(5), 548–562.
Zhao, G., Kingston, J., Kerremans, K., Coppens, F., Verlinden, R., Temmerman, R., & Meersman, R. (2004). *Engineering an ontology of financial securities fraud* (Lecture Notes in Computer Science) (pp. 605–620). Springer.

Chapter 8
What Does It Mean to Speak of a Model's Ontology? Term Structure Models and Their Migration from Financial Economics Into the OTC Derivatives Markets

Taylor Spears

Abstract This chapter examines the migration and translation of so-called "equilibrium" interest rate term structure models from academic financial economics into the over-the-counter (OTC) derivatives markets during the 1980s and 1990s. Initially developed to understand the economic factors that shape the pricing of bonds, these models underwent significant change to align with the modelling culture of derivatives practitioners as they migrated into the OTC derivatives markets. Using document analysis and interviews with quants and financial mathematicians, this study highlights the interplay between model stability and flexibility. It emphasizes that successful model migration involves iterative translations that maintain theoretical commensurability while adapting to new domains. The findings suggest that understanding the performativity of financial models requires examining the translation processes that enable models to reflect and integrate into the institutional contexts of their application. This case contributes to the Social Studies of Finance and the Philosophy of Science by demonstrating how model migration is facilitated by the translation of models into new forms that align with the organizational and practical realities of their target domains.

8.1 Introduction[1]

Within the social studies of finance (SSF) and adjacent fields, the study of financial modelling has been organized around several themes since the field's inception in the early 2000s. Early work on financial modelling was centered around Michel

[1] Portions of this chapter have been adapted from the author's PhD thesis: Spears (2014).

T. Spears (✉)
University of Edinburgh Business School, Edinburgh, UK
e-mail: taylor.spears@ed.ac.uk

E. Ippoliti et al. (eds.), *Philosophy and Finance: Ten Open Questions*, Synthese Library 504, https://doi.org/10.1007/978-3-031-96510-4_8

Callon's concept of the performativity of economics and sought to understand how financial models shape financial markets (cf. Beunza & Stark, 2004; Callon, 2007; MacKenzie & Millo, 2003). The concept of performativity later attracted criticism, both due to its tendency to underemphasize the importance of political and cultural factors in shaping finance (Mirowski & Nik-Khah, 2008; Riles, 2010), as well as for the tendency of strong versions of the performativity thesis to overstate the agency of models and other calculative devices (Felin & Foss, 2009; Marti & Gond, 2018). A second cluster of work focused instead on cultures of financial modelling, which has shed light on the diversity of modelling practices used in contemporary finance (cf. Hansen, 2020; MacKenzie & Spears, 2014; Mikes, 2009; Svetlova, 2018).

A third, emerging theme is the *translation* of models and calculative devices into new fields and organizations; that is, the process of aligning them to fit with the practices, infrastructures, and value systems of particular organizations and fields. The concept of translation, which was implicit in some of the early work on the performativity of models (MacKenzie, 2006), originates from Michel Serres's work and was later incorporated into actor-network theory and organization studies (Czarniawska & Sevon, 1996; Latour, 1984). It stands in contrast with the traditional concept of diffusion, which implicitly assumes that technologies, practices, and—in this case—models can migrate unaltered across firm and organizational boundaries. Translation studies, by contrast, take as a starting point the idea that the movement of ideas and practices usually entails the transformation of the object being translated. Developing a robust account of model translation is important for addressing certain deficiencies with the concept of performativity. Beunza and Ferraro (2018), for instance, show how it can overcome performativity's tendency to downplay the importance of political factors in shaping model use. Svetlova's (2014, 2018) detailed studies of financial modelling, meanwhile, have implicitly highlighted the translation work necessary in applying financial models in fields, such as finance, where models are evaluated according to commercial, rather than epistemic, criteria. In the financialization literature, financial models are seen as important carriers of financialized practices (Besedovsky, 2018; Chiapello & Walter, 2016; Coombs & van Der Heide, 2018). However, because models are altered as they move across organizations and fields, the capacity of a model to shape a particular domain depends on how that model is translated into that field or organization. Thus, understanding how models change during the translation process is crucial for comprehending the extent and manner in which financial models can exhibit performative effects.

To better understand the dynamics of model translation, researchers in the SSF have begun to combine the practice-oriented view of modelling characteristic of the SSF and the Social Studies of Science with a modular view found in the Philosophy of Science literature (van Der Heide, 2019). Practice-oriented accounts of modelling see modelling as a practice or activity, and emphasize that models change through a process of *bricolage*, a term borrowed from Lévi-Strauss (1968) to describe a process of creative tinkering (MacKenzie, 2003), by model developers or users. However, practice-oriented accounts of modelling are generally less specific about what changes when models are translated. By contrast, what van Der Heide (2019) calls the "modular" perspective provides several useful middle-range concepts through which to understand how models travel into and are adapted to particular domains. In his work on the adoption of *market consistent* valuation techniques from financial economics into insurance, he has shown how model translation can be studied through the lens of *model migration*. In doing so, he draws on a distinction made by Bradley and Thébault (2019) between model migration and model imperialism, as described by Mäki (2009). Model imperialism, which is used to describe cases such as Becker's (1990) application of economic models to understanding non-economic domains, such as the family, refers to a situation where an existing model's domain of application is enlarged to encompass new types of phenomena in another domain. By contrast, model migration occurs when an existing model is altered to operate within an entirely separate field or domain that lacks a shared ontology with the original field.

The existence of a shared ontology, according to this perspective, is thus an important differentiator between model imperialism and migration. If model imperialism implies ontological unity between a model's domain of origin and its target field, the notion of model migration instead suggests that models are adjusted to align with the ontology of a target domain through a process that Bradley and Thébault (2019:89) refer to as "re-sanctioning". As van Der Heide notes, however, this creates a puzzle for studies of modelling in the SSF: How can models exhibit performative effects when "migrating" to organizations and fields in the absence of a shared ontology with those domains? To resolve this puzzle, he argues that models embed particular ontological elements that remain stable even as they travel across domains, which serve as important sources of constraint that give those models agentic qualities. In doing so, van Der Heide draws on a notion of a modelling *ontology* from the SSF, a term used to denote assumptions embedded within a model and its related practices about "what economic value consists of and of the nature of the economic processes that create it" of particular models (MacKenzie, 2017). In van Der Heide's account, these ontological elements create what he calls "rough edges", or points of friction between the model and the target domain, which can be the source of the model's political effects.

van Der Heide's use of ontology follows a broader "ontological turn" in the social sciences (Jensen & Morita, 2016; Mol, 2003). Within the SSF, such work exhibits a focus on both the objects that constitute financial markets (Muniesa et al., 2011) as well as an examination of how particular objects are enacted through financial practices and infrastructures (Kennedy, 2017; Pinzur, 2021). This focus on ontology is also grounded in studies of *epistemic cultures*, wherein work tends to be organized around scientific objects (Knorr Cetina et al., 2001; Knorr-Cetina, 2000). In this chapter, I build on this work by examining the migration of interest rate term structure models from academic economics into the markets for derivatives that are traded on an *over-the-counter* basis during the 1980s and 1990s. Term structure models are used to simulate changes in the *yield curve*—the relationship between a bond's maturity date and its yield (i.e. return)—in a financially realistic manner. In particular, I focus on the translation of what are known as short rate models, which describe yield curve movements in terms of a small number of stochastic processes. These models were first developed by academic financial economists in the late 1970s but later became widely used by banks active in the markets for OTC derivatives for pricing and hedging so-called "exotic" interest rate derivatives that depend on the movement of multiple interest rates. The case I present here is based on a combination of document analysis of published papers and a series of 33 interviews conducted between 2011–2013 with derivatives quants active in the interest rate derivatives markets and financial mathematicians who were involved in the development of interest rate term structure models. My analysis in this chapter has also been informed by the practitioners' own accounts of the history of term structure modelling, most notably those of Rebonato (2004a, 2004b), who have also noted the increasing alignment between the mathematical structure of term structure models and the needs of complex derivatives traders.

The case that I present in this chapter combines the perspectives of both van Der Heide and Bradley and Thebault. Term structure models, I suggest, exhibited both ontological constraint *and* flexibility as they migrated between financial economics and the OTC derivatives markets. Indeed, it was this dual property—to enforce constraint along some dimensions and allow flexibility along others—that was critical for their successful migration into the OTC derivatives markets. On the one hand, elements of these models remained stable as they migrated from financial economics to the OTC derivatives markets, which generated affinities between these fields and allowed economists to work as "carriers" of these quantitative techniques into the OTC derivatives markets (Himick & Brivot, 2018). Both sets of models—for instance—employ the mathematics of stochastic calculus and make extensive use of the style of no-arbitrage reasoning associated with the Black-Scholes-Merton model. Both therefore shared a common modelling ontology in the sense of MacKenzie (2017: 38): a set of assumptions about "what economic value consists of and of the nature of the economic processes that create it".

At the same time, the equilibrium term structure models developed by academic financial economists could not be usefully applied to the problem of derivative pricing. Instead, as this chapter shows, the mathematical structure of these models was successively altered to fit within the distinctive modelling culture practiced by interest rate derivatives practitioners, which differed from that of financial economists in important ways. In this chapter, I focus on one such difference: the tendency of derivatives practitioners to make a conceptual separation between what are known as yield curves and implied volatility surfaces. This conceptual separation reflected (and to a significant degree, still reflects) the organization of derivatives risk management within dealer banks but was largely absent from early term structure models developed by financial economists. Quants, by contrast, needed models that could take prices from these two objects as model inputs. While both sets of term structure models had much in common, those that became popular within the derivatives markets were designed to "fit" market prices derived from these objects. As such, this chapter builds on a line of thinking developed by Svetlova and Dirksen (2014) and Svetlova (2014) that models are not only evaluated by epistemic criteria, but also by their pragmatic usefulness. I extend this finding by suggesting that these non-epistemic criteria may include the degree to which models reflect elements of the organizations in which they are used.

The remainder of this chapter is structured as follows. In the next section, I introduce term structure models and the objects they are designed to model, known as yield curves. Yield curves, I suggest, are complex objects in their own right, which organize work in both financial economics and in the financial markets. In section three, I then examine the emergence of the first short rate models in financial economics. In section four, I describe the modelling culture of interest rate derivatives practitioners, particularly the so-called "quants" who are responsible for developing and maintaining derivatives pricing models in dealer banks that are active in making-markets in these products. In section five, I examine how these models were translated into the working practices and infrastructures of the OTC derivatives markets. Finally, in section six, I conclude by discussing the relevance of this case for our understanding of modelling in the Philosophy of Science and the Social Studies of Finance.

8.2 The Yield Curve

This chapter's focus on the modelling culture of interest rate derivative practitioners takes as its starting point Knorr-Cetina's observation that in knowledge-intensive fields, particularly in science, the structure of work has become largely organized around what she calls *objects of knowledge* or *epistemic objects* (Knorr Cetina et al., 2001). Unlike everyday objects which function as black boxes, these objects are always changing and unfolding; they have multiple material instantiations but exhibit an "incompleteness of being" in that any particular instantiation only provides a limited view of the underlying object. In describing epistemic objects in this way, Knorr-Cetina had in mind scientific objects from fields such as physics and biology. However, this description also effectively describes what are known as *yield curves* in finance. The yield curve captures the relationship between the yield-to-maturity (i.e. return) of otherwise identical fixed-income instruments that vary only by their maturity date, i.e. the date at which a bond's principal amount will be repaid to investors. Yield curves are currency and instrument-specific representations of what economists and finance practitioners call the *term structure of interest rates*. They organize a significant amount of work within the financial markets, particularly the bond and fixed income markets, as well as certain strands of work in academic financial economics.

As an epistemic object, the yield curve has its origin with developments in central bank policy and the quantification of the UK bond markets during the 1950s–1970s, but its use among practitioners became widespread with the widespread adoption of Bloomberg Terminals in the 1980s (Cassar, 2022). Fixed income practitioners came to "read" changes in the yield curve as a sign of the market's expectation of future economic developments (Zaloom, 2009). This is because of how the yield curve moves in such a way that can shed light on market participants' changing expectations around interest rates. In general, yield curves tend to slope upward, reflecting the tendency of bond investors who are willing to lend money over longer periods of time to be rewarded with a higher return. But this is not always the case. For example, when market participants expect that a recession is imminent, they will often sell shorter maturity bonds and buy longer maturity ones, depressing the yield of the latter while raising the yield of the former. This is known as an *inverted* yield curve. In general, yield curve movements exhibit considerable complexity; they may shift, rotate, and even twist. But their movements are not wholly random. Points along the yield curve often move together; their movements are thus correlated.

Yield curves are not "natural" objects explicitly visible in the bond markets; instead, they must be constructed from the quoted prices of bonds and other fixed-income instruments, such as swaps. Yield curves thus exhibit ontological multiplicity. Indeed, there is no *one* yield curve. Yield curve construction is an inherently theory-laden exercise, as a yield curve for a particular market is necessarily underdetermined by observed prices of traded fixed-income instruments. Indeed, while the mathematical description of a yield curve assumes that fixed income instruments are traded with maturities that span a continuum of future maturities, in reality there is only ever a handful of maturity dates for which bonds trade at an acceptable level of liquidity to derive meaningful information about yields. To construct a yield curve, one must *interpolate* between observed bond yields, which entails making assumptions about how bond prices behave for unobserved maturities. Not surprisingly, different practitioner communities in finance thus construct yield curves in distinct ways. Central bankers that forecast changes in the yield curve often require parsimonious models of the yield curve that relate its movements to specific macro-economic variables (Cassar, 2022). Among derivatives practitioners, consistency with the assumption of no arbitrage is paramount: the idea that it should be impossible to make a profit through trading without taking any risk (Hagan & West, 2006). If this assumption is not met, a trader cannot expect to earn a positive profit when using a model to price and hedge derivatives.

Yield curves, technically known as *zero-coupon curves*, can also be translated into alternative forms that are more useful for certain tasks. Using a mathematical formula, a yield curve can be converted into a *discount curve*, which for each maturity date provides the price of a hypothetical bond that pays one unit of currency at some future date and nothing before then. These discount bond prices can then be used to compute the present value of future cashflows, a critical step in any financial valuation exercise. If one grants the assumption that bond markets are free from opportunities for risk-free arbitrage, then one can also derive what are known as *forward rates*, represented via a *forward curve*, which represent the market's expectation of the cost to borrow at some future date rather than today. For practitioners, forward rates are useful in that they allow one to answer questions such as, "How much should one pay today for a financial contract that will allow one to borrow at the one-year interest rate obtaining in six months' time?" They are also critical inputs when valuing certain cashflows, particularly those that involve floating (i.e. non-fixed) interest rates. Forward rates can, likewise, be expressed in different forms. While practitioners generally prefer *discrete forward rates*, which align conceptually with the rate specified in a derivative contract, known as a forward rate agreement, to borrow money over a discrete time period, mathematicians often use *instantaneous forward rates*, which represents the market-implied cost to borrow over an instantaneous period of time at some future date. Yield curves can thus be expressed using different conceptual and material instantiations, each of which highlights certain features of interest rates over others. Like many epistemic objects, they thus lack a "completeness of being" (Knorr Cetina et al., 2001:181) and working with them necessarily involves a constant process of translation from one form to another.

In certain circumstances, forecasting or simulating future changes in a yield curve, discount curve, or forward curve is necessary. In this case, an *interest rate term structure model* is needed to simulate its movement in a way that is financially realistic. The users of term structure models are varied. Central banks use such models to forecast the impact of monetary policy on the economy (Cassar, 2022). Likewise, bond traders will typically use term structure models to identify mispricing between bonds of different maturities. One of the more prominent groups of users of these models in the 1990s and 2000s were derivative practitioners, who used term structure models to price and hedge the risks associated with buying and selling 'exotic' derivatives that depend on the movement of multiple interest rates. However, term structure models originated within financial economics beginning in the late 1970s. Before examining the migration and translation of these models into the derivatives markets, we therefore first examine their initial development in academic financial economics.

8.3 Term Structure Modelling in Financial Economics

In the mid-twentieth century, the shape and behavior of the yield curve became a subject of inquiry within economics. Models that economists developed to explain the yield curve's empirical behavior reflected the discipline-specific ontology of economics, particularly the notion that investor preferences are a primary determinant of asset prices. Indeed, a key focus of economists came to be the existence and nature of the *bond risk premium*: the notion that the yield curve's upward-sloping tendency can be explained by the fact that holders of long-term debt demand compensation for the risks inherent in these instruments (Culbertson, 1957; Hicks, 1946; Modigliani & Sutch, 1966; Roll, 1970). However, the empirical behavior of this premium was unclear. So-called "equilibrium" term structure models were developed, in large part, to better understand the economic factors that shape the market's valuation of bonds. The first such model was developed by Oldřich Vašíček, a Czech mathematician in the late 1970s (Vašícek, 1977).

Vašíček employed the mathematical technology of derivatives pricing pioneered by Black, Scholes, and Merton to understand the equilibrium pricing of bonds. His work was based on the following analogy: if an option can be thought of as a derivative on a particular share of stock, could the same mathematical techniques be applied to the problem of bond pricing by thinking of bonds as derivatives of interest rates? In working through this analogy, Vašíček faced a difficulty. In the Black-Scholes-Merton approach, a stochastic differential equation defines the evolution of a single stock's price. It was not clear how one could apply this approach to an entire yield curve, made up of multiple rates. The solution Vašíček developed was to find "the smallest common denominator, so to speak, for [bonds of] all maturities", and to treat that as a fundamental building block that drives the random movement of bonds along the whole yield curve (Vašícek). In Vašíček's model, this is the *short rate*, which represents the continuously compounded interest earned on making a risk-free deposit into a bank account and is denoted r_t. Thus, the starting point of the Vašícek model is what is now called the *money market account*, whose change at each moment in time is defined by the following differential equation:

$$dN(t) = N(t)\, r(t)\, dt$$

where r_t denotes the current value of the short rate and N_t represents the current value of the account. Using this specification of the money market account and a description of the dynamics of the short rate, one could work out prices for zero-coupon bonds. To this end, he provided a worked-out example that showed how one could calculate prices for bonds assuming that the short rate follows a mean-reverting stochastic process that mathematicians call an Ornstein-Uhlenbeck process. This property of mean-reversion captures the real-world tendency of interest rates to tend towards a long-run average. With this process, interest rate movements in each time period are composed of a deterministic component (the drift term) and a random component, which is made up of a Gaussian random variable scaled by a parameter denoted by σ. The drift term is made up of several parameters that capture empirical properties of the yield curve: a long-run average level for the short rate, denoted by θ, and a parameter which captures the speed at which the short rate returns to this long-run average, denoted by α. In the stochastic differential equation defining the short rate below, r_M denotes the current market short rate:

$$dr(t) = \alpha\,(\theta - r(t))\, dt + \sigma dW(t)$$

$$r(0) = r_M$$

Vašíček's model can be solved for the price of a bond using what is known as *risk-neutral valuation*, a technique central to the Black-Scholes paradigm for pricing derivatives based on the idea that if a derivative is fully hedged then the derivative and its replicating portfolio must, in equilibrium, earn the risk-free rate of interest (Cox & Ross, 1976; Harrison & Kreps, 1979). To apply this idea, one must switch from the so-called "real-world" probability measure, which describes the empirical behavior of the short rate, to a "risk-neutral" probability measure where discount bond prices exhibit what is known as the *martingale property*, a situation where bond prices are free from arbitrage and therefore the expected value of the price remains the same from one period to the next. This change of probability measure is accompanied by a change to the drift of the short rate process. When this drift adjustment is made, a new parameter (typically denoted by λ) must be added, which captures what is known as the *market price of risk*, a measure of the extra compensation that investors demand for taking on interest rate risk:

$$dr(t) = [\alpha(\theta - r(t)) + \lambda\sigma]\,dt + \sigma dW(t)$$

Using this short rate equation, one can then derive a formula for the price at time t of a zero-coupon bond maturing at time T, denoted by $P(t, T)$:

$$P(t, T) = e^{A(t,T) - B(t,T)r(t)}$$

$$B(t, T) = \frac{1}{\alpha}\left(1 - e^{-\alpha(T-t)}\right)$$

$$A(t, T) = \left(\theta + \frac{\lambda\sigma}{\alpha} - \frac{\sigma^2}{2\theta^2}\right)(B(t, T) - (T - t)) - \frac{\sigma^2}{4\alpha}B^2(t, T)$$

For our purposes, the important thing to note about these formulas is that they express bond prices as a function of a small number of parameters—α, θ, λ, and σ—representing real-world behavior of the yield curve and investor preferences. These parameters can, in principle, be estimated via statistical methods from historical data on interest rate movements. Having derived a set of bond prices from the model, one could then—again, in principle—price derivatives on those bonds.

Vašíček's short rate approach was extended by other financial economists and practitioners with the aim of better understanding the economic factors that shape the pricing of bonds. These models were also designed to derive bond prices consistent with the assumption of no arbitrage from a small number of parameters that captured important features of the "real-world" behavior of the yield curve in order to understand how investors' risk preferences shaped the equilibrium pricing of bonds. For example, John Cox, Jonathan Ingersoll, and Stephen Ross, together independently developed a short rate model similar to Vašíček's but where the short rate equation is drawn from a general equilibrium model of the economy to ensure its consistency with economic theory (Cox et al., 1985). Michael Brennan and Eduardo Schwartz, by contrast, developed a term structure model in which the yield curve is driven by both changes in the short rate and a second factor capturing long-term interest rates (Brennan & Schwartz, 1979). Their model addresses a shortcoming of Vašícek's model, which implicitly assumes that yield curves only 'shift' up and down, in contrast to their much more complicated empirical behavior. Variants of these models were later adopted by bond traders, particularly arbitrageurs, due to their ability to help one identify mispriced bonds (Cassar, 2022; Rebonato, 2004a; Tuckman, 2022).

8.4 The Modelling Culture of Interest Rate Derivatives Practitioners

Following the growth of an options market centered around the Chicago Board Options Exchange (CBOE) in the 1970s, a combination of factors led commercial and investment banks to begin functioning as dealers in "over-the-counter" options on bonds like US Treasuries, as well as options on the London Interbank Offered Rate (Libor). Demand for these products was initially driven by both an overall decline in interest rates from their historical highs in the late 1970s, as well as an increase in interest rate volatility, which had two effects. On the one hand, 'yield hungry' investors began to seek out new sources of returns; on the other hand, a broad range of corporate actors and market participants became increasingly exposed to changes in market interest rates and thus sought new techniques for managing these risks.

To price and hedge many of the so-called "exotic" interest rate derivatives that became popular, a term structure model is needed to model the simultaneous movement of multiple points along the yield curve in a manner consistent with the assumption of no arbitrage. And while equilibrium term structure models, such as Vašíček's, could be used to derive theoretical prices for these instruments, these models were never widely adopted for pricing interest rate derivatives. The reason for this was not primarily because of their predictive performance but because they failed to align with elements of the modelling culture that organized interest rate derivatives trading activities within dealers.

Central to this culture is a modelling practice known as "market-implied" calibration (Rebonato, 2004b; Wilmott, 2000). This refers to the practice of "backing out" the parameters of a model from market prices in such a way that the model is capable of "reproducing" those quoted prices (Svetlova, 2018). In more technical terms, calibrating a model involves solving a particular optimization problem: that of finding values for the model's parameters that minimize the difference between the actual and predicted prices of a chosen set of calibration instruments. Once calibrated, a model is then used to price derivatives with uncertain values, namely exotic derivatives and vanilla derivatives whose prices are not actively quoted. Paul, an interest rate quant interviewed in 2012, explained the process of interest rate model calibration in the following terms:

> Now, a model right out of the box is completely useless. The reason is because it has mathematical parameters, and until you set those parameters you aren't going to be able to price anything. So the next step is always calibration. And when you calibrate in the interest-rate world, you have to do two things: first is you've got to get the discount curve correct, because you are going to be toast otherwise. The second thing is typically you select a set of vanilla instruments - swaptions, caps, floors - things that you know moment-

> by-moment throughout the day what their prices are. And you make sure that your model can reproduce market prices on the calibration set.

As Paul suggests, term structure models used in this context must be capable of "get[ting] the discount curve correct" as well as "reproducing" market prices for a set of instruments chosen to hedge the derivative. This requirement arises from the fact that exotics traders hedge their positions using simpler interest rate derivatives, namely interest rate swaps, caps, and swaptions. If a term structure model were incapable of matching these prices, the model would not produce accurate information about how to hedge a given 'exotic' structure. Moreover, because there is very low liquidity for exotic derivatives, these vanilla instruments are used to *value* the derivative itself in accordance with generally accepted accounting principles (IFRS Foundation, 2013). Using a model that is incapable of matching these prices would thus almost certainly create a loss for the trader. As Aaron, another quant, explained in an interview in 2012:

> What is absolutely essential to the trader is that the day he does the trade he is on market. Why? Because he will hedge him or herself by doing plain-vanilla trades in swaptions and caplets. So if the model does not predict the correct prices for the hedges, he will record a day-one discrepancy, which will almost invariably be a loss. So that is a must.

Another important feature of this modelling culture is a conceptual distinction between the yield curve itself—what Paul above calls the discount curve—and what are known as *implied volatility surfaces*, which are used to represent the prices of interest rate options like caplets and swaptions. Implied volatilities are computed by adjusting the volatility parameter σ in the Black-Scholes model until the model's price matches the observed market price of the option. The practice therefore establishes a connection between the market's expectation of the future volatility of interest rates and the prices of interest rate options. In the case of swaptions, the volatility surface can be thought of as a series of matrices, for which one axis corresponds to the swaption's expiry date and the other axis corresponds to the tenor of the underlying swap. The most important of these are known as "at-the-money" swaption volatilities and are most often used to hedge interest rate exotics. Within dealer banks, separate trading desks typically handle market-making for different types of interest rate derivatives: one desk for swaps and other linear derivatives, and another for interest rate options such as caps, swaptions, and floors. Each desk is responsible for "publishing" prices for these instruments, which exotics traders take as inputs to their models. The linear products desk produces publishes these prices in the form of discount and forward curves, while the options desk generates volatility surfaces. As a consequence, term structure models were evaluated by derivatives practitioners to a significant degree by their capacity to 'match' these existing prices and reflect the conceptual separation between these two sets of objects, a need that is tied to the division of labour on the trading floor between different trading desks.

As Fig. 8.1 illustrates, the Vašíček model faces difficulty in even matching an observed yield curve. As noted previously, early equilibrium models, including the one developed by Vašíček, had only a small number of parameters. This was a justifiable choice in the context of the modelling culture of academic economists:

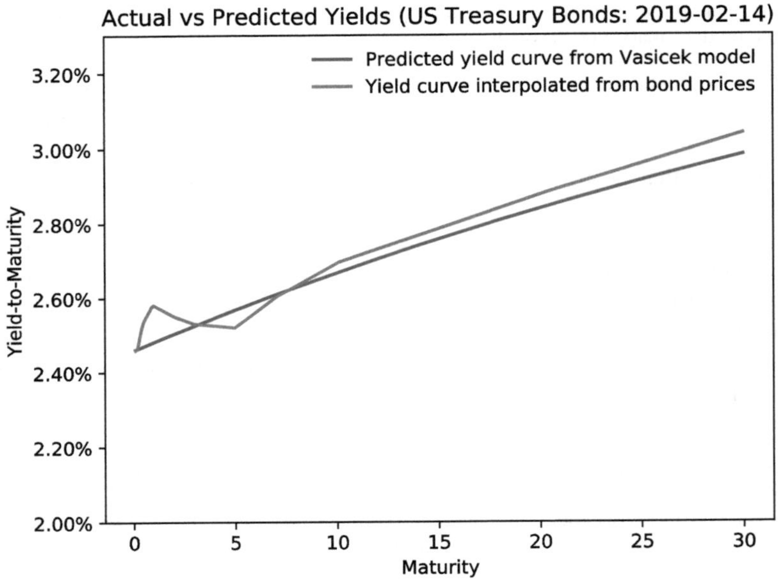

Fig. 8.1 Comparison of predicted and actual yield curve for a Vašícek model (Least squares calibration was performed to US Treasury yields on 2019-02-15 for 1M, 2M, 3M, 6M, 1Y, 2Y, 3Y, 5Y, 10Y, 20Y, 30Y maturities using SQP optimization via scipy.optimize.minimize. The yield curve was constructed using the ZeroCurve class in QuantLib. Underlying bond yield data was sourced through Quandl)

as Mary Morgan (2012) has shown in her work on the history of economic modelling, models that are created and used by economists are designed to be "small worlds" that contain "compressed" descriptions of real-world economic processes. Moreover, as Rebonato (2004a) explains, this property is precisely why equilibrium models could be useful to certain financial users, like relative value bond traders, who use these models to identify mispriced bonds. However, equilibrium term structure models had insufficient degrees of freedom to calibrate either to an existing yield curve or to the prices of interest rate options. For derivatives practitioners, this represented a critical shortcoming. Emanuel Derman, a well-known quant, faced this problem in the late 1980s as he, along with Fischer Black and Bill Toy, developed a term structure model that could be used to price and hedge bond options. Derman ultimately co-developed what is known as the Black-Derman-Toy (BDT) model, a term structure model that became popular in the 1990s (Black et al., 1990). In an interview conducted in 2012, he highlighted the importance of compatibility with market-implied calibration as a necessary constraint on the BDT model:

> Vašíček's [model] specifies the dynamics for interest rates, and calculates the yield curve. And then you can calculate options on the yield curve. But what we were interested in was fitting the yield curve and then calculating options on the yield curve. And if you took the simple version of Vašíček, you ended up with bond prices which weren't the bond prices in the market. [...] The aim was: give me a yield curve; I'll fit it, and then I'll calculate

> option prices. Because it was sort-of pointless to calculate the value of an option on a bond when you had already priced the bond incorrectly. So, we were doing something similar to Vašíček, in a way, but our aim wasn't to predict the yield curve, but to calibrate to the yield curve.

In the next section, I examine how term structure models were translated to accommodate this modelling ontology.

8.5 Model Migration and Translation

Throughout the 1980s and 1990s, various economists and practitioners became involved in the development of alternative term structure models that fit within this modelling culture. In some cases, these involved ad-hoc modifications to equilibrium term structure models, like the Vašíček model. As one quant explained in an interview, many practitioners coming from a physics background were comfortable making modifications to "off-the-shelf" term structure models on their own. Records of these ad-hoc translation efforts are not publicly observable. However, several important term structure models aligned to the needs of derivatives practitioners were published in academic and practitioner journals.

One of the first was the Ho-Lee model, which was developed by Tom Ho, a financial economist at New York University's Business School, along with his graduate student, Sang-Bin Lee. Ho had decided to work on the problem of pricing interest rate options after a visit to Merrill Lynch's trading floor led him to see that there was a need for better tools and models on the fixed-income side (Ho Interview). To acquaint himself with the literature on term structure modelling, Ho had studied Cox, Ingersoll, and Ross's paper on term structure modelling. After doing so, he and Lee developed a model which, in contrast to the Vašíček and Cox-Ingersoll-Ross models with their focus on the short rate, modelled the movement of the entire yield curve simultaneously. As Ho explained in an interview in 2012:

> There's no reason why you should look at only one rate, because we know the whole yield curve is moving. And surely, a general equilibrium model [such as the CIR model] cannot explain how the shape of a yield curve moves - it's too complicated. So the only way to do it, really, is take it the yield curve as a given (Interview with Ho).

To build a term structure model capable of taking a yield curve "as given", Ho and Lee adopted a modelling technique outside of term structure theory: the Cox-Ross-Rubinstein (CRR) binomial options pricing model. While still grounded in the no-arbitrage logic of Black-Scholes, the CRR model assumes that the price of the underlying asset can only move up or down by a fixed amount in each time period. Ho and Lee adopted this approach to model the entire discount curve as evolving either up or down by a fixed amount in each time period, in contrast to the Vašíček model, in which a single point—the short rate—evolves and bonds are priced as a function of this short rate. This allows the Ho-Lee model to "reproduce" an observed yield curve with ease, as is required by interest rate derivative practitioners. Ho explained in the same interview that he was aware his approach was deviating from the modelling practices characteristic of economics. Ho noted that his colleagues at NYU were puzzled by—and even dismissive of—his work as it lacked many of the characteristics of economics' modelling ontology.

> I could not explain it to economists. Because to understand, "why do you have all of these equations? What are you actually looking for? Where's the supply and demand curve? Where's the preferences? Where's the risk aversion?" None of that is in my equation (Interview with Tom Ho).

While not well-aligned to the modelling culture of academic economics, Ho and Lee's model instead aligned more closely with the modelling culture of interest rate derivative practitioners given that it takes the market yield curve "as given" and drops the focus on investor preferences characteristic of equilibrium term structure models. As I explained in the previous section, because interest rate options traders hedged their positions by trading in bonds, it was critically important that pricing models take *as inputs* the prices of these instruments via the yield curve. Ho and Lee's model was a translation in a further sense: in contrast to equilibrium term structure models, which assume that time is a continuous quantity, the original Ho-Lee model was a discrete time binomial model. While limited in its applicability, particularly for pricing exotic derivatives, the model was initially very attractive to practitioners due to its binomial structure, which allowed it to be implemented on desktop spreadsheet applications.

While useful for pricing bond options, it was limited in its usefulness for pricing exotic interest rate derivatives. It could calibrate to an observed yield curve, but the Ho-Lee model assumes that interest rate volatility is constant as a function of maturity. As a consequence, it struggles to "reproduce" the quoted prices of interest rate options, a necessary feature for a model used to price exotic derivatives dependent on multiple interest rates. Moreover, the discrete time formulation of the Ho-Lee model made comparing it to existing term structure models difficult. In the years following its publication, both derivatives practitioners and academics independently made further translations of the model, in part to allow it to calibrate to a wider range of financial instruments. At Merrill Lynch, a quant named Farshid Jamshidian showed that the Ho-Lee model could alternatively be represented in the form of a continuous-time short rate model, akin to the Vašíček model, where the drift parameter is time dependent (Jamshidian, 1996).

$$dr_t = \theta(t)dt + \sigma d\mathrm{W}\,(\mathrm{t})$$

$$\mathrm{r}(0) = r_M$$

Translating the Ho-Lee model back into a continuous-time formulation helped quants to better understand its relationship with short rate models, such as Vašíček's. Jamshidian (1996) also showed how mean reversion could then be added back into the model, yielding a mean-reverting model similar to Vašíček's but with a time dependent drift. Independently from Jamshidian, John Hull and Alan White—two finance professors at Toronto with close ties to the growing derivatives industry—

developed an extension of Vašíček's model that allows for both time-dependent drift and volatility parameters to achieve an exact fit to an observed yield curve and a selection of interest rate options. Hull and White also developed numerical techniques for using the model to price derivatives (Hull & White, 1994a, 1994b). In one version of their model, the short rate instead takes the following form:

$$dr(t) = (\theta(t) - a(t)\mathrm{r}\,(\mathrm{t}))\,dt + \sigma\,(\mathrm{t})\,dW(t)$$

$$r(0) = r_M$$

Described by Hull and White as an Extended Vašícek model, interest rates in it follow a "mean-reverting" stochastic process, almost identical to the Ornstein-Uhlenbeck process used in the Vašíček model. However, through the addition of time-dependent drift parameters, denoted by $\theta(t)$ and $a(t)$, it can be calibrated to exactly match to an observed yield curve. Time-dependent volatility parameters, denoted by $\sigma(t)$, allow the model to fit the prices of interest rate options. Figure 8.2 shows the result of calibrating a Hull-White model to an observed yield curve. In the figure, the predicted and actual yields exactly align, which is why there only appears to be a single yield curve. Table 8.1, by contrast, illustrates the model's capacity to "match" observed swaption prices when calibrating to the 'diagonal' entries of what

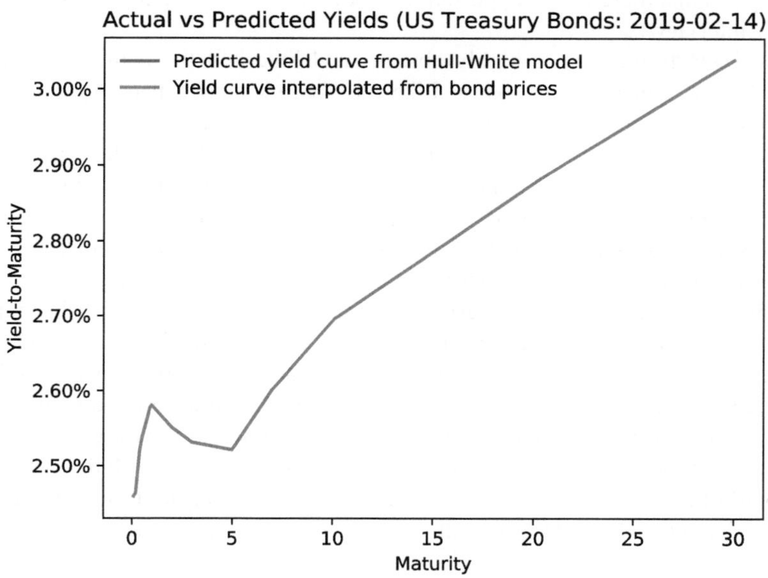

Fig. 8.2 Comparison of observed vs. predicted yield curve for a Hull-White model (Calibration was performed via the HullWhite class in the QuantLib library in Python. The underlying yield curve was constructed via the ZeroCurve class in QuantLib. Interest rate data was sourced through Quandl.)

Table 8.1 Relative error of model-calculated swaption prices using a Hull-White model when calibrating to the 'diagonal' of the at-money swaption volatility surface[a]

Exp/Tenor	1	2	3	4	5
1	126.70%	59.74%	33.97%	16.76%	**7.14%**
2	36.39%	17.64%	7.38%	**2.09%**	−2.74%
3	9.73%	1.20%	**−1.76%**	−5.19%	−4.20%
4	−4.58%	**−4.96%**	−7.36%	−5.10%	−4.32%
5	**−3.49%**	−6.01%	−2.76%	−1.77%	−1.68%

[a]Valuation performed using HullWhite class in QuantLib using JamshidianSwaptionEngine with USD swaption data from 2017-01-03. Yield curve constructed via ZeroCurve class in QuantLib. Data sourced through Quandl.

is known as the at-the-money swaption grid. As the table suggests, the degree of fit is generally good but not perfect.

Term structure models were further aligned with derivatives practitioners' modelling culture with the development of what is known as the Heath-Jarrow-Morton framework. Co-developed by David Heath, Robert Jarrow, and Andrew Morton at Cornell University following their efforts to implement the Ho-Lee model for a commercial client, the framework specifies the necessary and sufficient requirements that term structure models must satisfy to be compatible with the assumption that there are no opportunities for risk-free arbitrage (Heath et al., 1992; Jarrow interview).

The HJM framework is highly abstract and inherited the complexity of both the short rate and Ho-Lee models. Like the Vašíček model, it is expressed using continuous-time stochastic processes, rather than being a discrete-time binomial model. However, rather than modelling the movement of a single point (i.e. the short rate), the framework followed Ho-Lee by modelling the evolution of the *entire* yield curve simultaneously. It did so, however, in a new way by modelling the movement of what is known as the instantaneous forward interest rate curve $f(t, T)$ rather than the discount curve itself. The instantaneous forward rate curve, by contrast, represents the interest rate, set at date t, to borrow at a later date T for an instantaneously short period of time. It can be thought of as an idealized mathematical representation of the interest rate one would pay to lock-in today an overnight loan at a future date. The core of the HJM framework is a restriction on the drift term within the equation describing the evolution of the forward rate curve, which is necessary to keep forward rate movements consistent with the assumption of no arbitrage. When this drift restriction is incorporated, the following equation can be derived, which describes the arbitrage-free evolution of the forward rate curve:

$$df(t, T) = \sigma(t, T)\left(\left(\int_t^T \sigma(t, u)\,du\right)dt + dW(t)\right)$$

$$f(0, T) = f_M$$

In the above, $f(0, T)$ represents the current observed forward rate curve, which can be derived from the current market yield curve, while $f(t, T)$ denotes the curve at some other time in the future. The function $\sigma(t, T)$ represents the volatility of $f(t, T)$. It is a function that takes two inputs and can thus be thought of as a matrix with the starting date of the forward contract (denoted by t) as rows and its maturity date as columns (denoted by T).

The HJM framework is significantly more abstract than the Vašíček, Ho-Lee, or Hull-White models. Rather than modelling the movement of a single point through time, the forward rate curve is instead a continuous three-dimensional object, containing—in the most general case—an infinite number of forward rates. Given its highly general structure, the core of the framework—the HJM volatility structure, denoted by $\sigma(t, T)$—was left unspecified in the original paper. This level of abstraction initially made the framework difficult to implement, particularly using computing technology available within banks in the early 1990s. As Oren Cheyette (2002), a quant practitioner active in the markets for mortgage-backed securities in the 1990s, noted in an overview on interest rate modelling:

> [A] generic HJM model requires keeping track of a potentially enormous amount of information. The HJM framework imposes no structure other than the requirement of no-arbitrage on the dynamics of the term structure. Each forward rate of fixed maturity evolves separately, so that one must keep track of each one separately. Since there are an infinite number of distinct forward rates, this can be difficult (Cheyette, 2002: 12)

Adding to these difficulties is the fact that interest rates in the HJM framework are, in the general case, *non-Markovian*: in the model, the future level and shape of the forward rate curve depends not only on its current level and shape but the whole history of its movements up until the present time, thus greatly expanding the memory and computational requirements needed to implement the framework.[2] Making the HJM framework usable in practice was thus an area of applied research during the early 1990s within financial institutions involved in the interest rate derivatives markets (Rebonato, 2002). According to Jarrow, he, Heath, and Morton spent around 5 years "trying to convince everyone that you could compute with this", an effort that eventually led Andrew Morton to implement the model for Lehman Brothers and later run its fixed-income trading group (Jarrow and Morton interviews). Cheyette (1992) himself developed a popular modification to the HJM framework that ameliorates these computational difficulties by restricting the volatility structure $\sigma(t, T)$ to a "separable" function consisting of a time and maturity-dependent factors. Variants of this model are still used within certain dealers for pricing and hedging various "exotic" interest rate derivatives.

[2] This path dependence is created by the presence of $\int_0^t \sigma(u, t)$ in the equation defining the value of the forward rate, which indicates that the forward rate $f(t, T)$ depends on the complete history of the curve's evolution from 0 to t.

Despite its practical complexities, our interviews indicate that what made HJM initially appealing to derivatives practitioners was its mathematical structure, which aligned closely with the ontology implicit in quants' modelling culture. In the equation describing the evolution of forward rates above, the future value of a forward rate is specified entirely by two objects: the current forward rate curve $f(0, T)$ and the volatility surface $\sigma(t, T)$. This separation of the forward curve from the volatility surface corresponded with the conceptual separation that derivatives practitioners make between yield curves and implied volatility surfaces, which itself is rooted ins the organizational structure of the derivatives trading floor. Not only did this make it easier to understand how changes in the prices of hedging instruments might affect a derivative's value, but it made the problem of calibrating the model to these prices conceptually simpler. As Ryan, a quant who joined the industry in 1997, noted in an interview in 2012:

> The whole HJM approach was nice conceptually to a lot of us at the time, because it had the key understanding of separating out the forward curve from the volatility surface. In short-rate modelling those two were all mixed up together. Short-rate models can be as powerful as HJM but there are differences in understanding you get. With a short-rate model, the forward curve – you know, where the forward Libors are – is all mixed-up with the volatility surface and mean-reversion – it's all tangled. Whereas with the HJM it's much cleaner. You've kind-of got the forward curve in one place, and you've got the volatility surface in another. [...] So the idea that your rates model is based on the forwards, the vols, and the correlation as three separate planks. It's quite a nice intuitive idea because that corresponds a lot to the way that derivatives are priced in practice. So, the first thing you ask when you price derivatives is: "what's the forward?" And if you're doing an option "What's the vol?" And as soon as two or more things are involved, "What's the correlation structure?" And you often risk manage those in different ways, maybe different people are risk managing them. So the ability to separate them out is very useful. So the HJM was a good conceptual move forward in that respect because it achieved that separation.

As I explained in the previous section, on the trading floor of a dealer making markets in interest rate derivatives during this time, there was an organizational division between the trading desk that "makes markets" in linear interest rate derivatives, such as swaps and forward rate agreements, and the options trading desk which trades "volatility" in the form of caps and swaptions. In short rate models, such as the Hull-White model, the volatility surface and the forward curve are "all tangled", making it more difficult to understand to understand the risk factors affecting the valuation of a given derivative. Along with its ontological alignment with the practical organization of derivatives trading, the HJM framework also helped to organize quants' knowledge about term structure modelling itself. By the early 1990s, a variety of short rate models had been developed and used by derivatives practitioners, including the Hull-White model discussed previously. HJM helped to elucidate the mathematical relationship between these models, given that—as practitioners later learned—all arbitrage-free short rate models based on the mathematics of Brownian motions can be expressed using the HJM framework by choosing particular functional forms for the volatility structure (Baxter, 1997).[3] For this reason, in the present day, the primary use of HJM seems to be the construction of simpler models which are written in a form that aligns more closely with the modelling culture of derivatives practitioners than earlier short rate models. One such model, known as the Linear Gauss Markov (LGM) model, was introduced by Hagan and Woodward (1999). As one interviewee explained, it is "exactly Hull-White written like an HJM model" but is written in a form that allows for both automatic calibration to the yield curve and faster numerical performance:

> Interviewee: LGM is the Hull-White written this way - Linear Gauss Markov. And it's exactly Hull-White written like an HJM model, actually.
>
> Spears: So, instead of the instantaneous short rate you have an instantaneous forward rate?
>
> Interviewee: Yeah. And again, it is exactly the same model, because I can solve this one and get this, or start from here and get that. So, it has exactly the same range, if you know what I mean. It's just that this one - written this way - it's just a lot easier to deal with. It's just trivial.

8.6 Discussion and Conclusion

In this chapter, I have presented one historical episode in a broader history of the migration of term structure models from academic financial economics into the OTC derivatives markets. In moving from economics into the derivatives markets, term structure models underwent a series of translations to more closely align with the modelling culture of interest rate derivatives practitioners. One important aspect of this culture, which has been my focus in this chapter, is the conceptual separation between forward curves and implied volatility surfaces and the requirement that prices embedded in these objects be taken "as given" by models. These requirements, as I noted, are rooted in how models are used to value and hedge "exotic" derivatives and how derivatives trading is organized within investment banks. Term structure models from financial economics were not well-aligned with this modelling culture, as their aim was to predict the shape and level of the yield curve based on investor preferences and to understand the more general economic factors that shape bond prices. The resulting process of model translation involved both ad-hoc bricolage to existing models and the development of entirely new models that aligned with the ontology of this culture, such as the HJM framework. With its focus on the relationship between model migration, translation, and ontology, this historical episode sheds light on several aspects of mathematical modelling relevant to researchers in both the SSF and the Philosophy of Science.

---

[3] For instance, Minnis (2013) shows that the Ho-Lee model corresponds to a choice of $\sigma(t, T) = \sigma$ while the Hull-White model corresponds to a choice of $\sigma(t, T) = \sigma e^{-a(T-t)}$.

First, this episode suggests that models which successfully migrate across fields or organizations exhibit a combination of ontological flexibility *and* stability. In their account of model migration, Bradley and Thébault (2019) emphasize that when models migrate across fields, they must undergo a process of "re-sanctioning" through which they are modified and re-justified according to the epistemic norms of the target field or domain. Their focus is thus on the changes that models undergo when moving between domains. By contrast, in his work on the migration of risk-neutral valuation models from the derivatives markets into insurance, van Der Heide (2019) instead focusses on the ontological elements that remain stable even as these models migrated into the insurance field. It is these stable elements, according to van Der Heide, that are the source of these models' political effects on this field.

The models that I examine in this chapter instead exhibited both ontological change *and* stability. This is apparent if we compare the Vašíček model to the HJM framework. Both make use of the same underlying mathematics, namely the stochastic calculus of Brownian motions. Likewise, both are grounded in the logic of "risk-neutral" valuation associated with the Black-Scholes-Merton model, which is described by van Der Heide and MacKenzie and Spears (2014). These broad similarities are reminiscent of shared theoretical "templates" described by Knuuttila and Loettgers (2014) and "computational templates" of Humphreys (2002). However, the ontological structure of these models is significantly different: they model fundamentally different phenomena at different levels of analysis. Whereas the Vašíček model describes the movement of a single interest rate—the short rate—using a set of parameters that capture investor preferences and the "real-world" behavior of interest rates in order to derive the equilibrium pricing of bonds, the HJM framework describes the movement of the instantaneous forward rate curve in terms of an exogenously specified three-dimensional volatility structure with the aim of pricing interest rate derivatives. It was this property of the HJM framework that made it so appealing to derivatives practitioners, according to my interviewees, due to its alignment both to the practical organization of derivatives trading within dealer banks and its usefulness in elucidating the relationship between existing term structure models. Thus rather than conceptualizing of a model's ontology as *either* a fixed entity or as completely flexible, it is perhaps more useful to think of models as embedded in a flexible web of assumptions, "templates", conventions, objects, and associated practices. My analysis suggests that in the sequence of translations that helped to align term structure models to the modelling culture of interest rate derivatives practitioners, distinct elements of this ontological web remained stable at particular stages, leaving sufficient flexibility for the translation of these models into new domains while preserving theoretical commensurability. The HJM framework maintained the Ho-Lee model's focus on modelling the entire term structure instead of the short rate but focused on the instantaneous forward rate curve while adopting a continuous-time formulation. The Hull-White model, by contrast, maintained the general structure of the Vašíček short rate equation, but allowed for time-dependent parameters that accommodate the fitting of an observed yield curve and the volatility surface. Term structure models' capacity to migrate into new domains ultimately lay with the capacity of this web to flexibly accommodate institutional contexts different from those within which these models were originally developed.

Second, the account I develop in this chapter suggests that there is value in conceptualizing model migration as an iterative process that involves the development of sequences of new models, rather than the migration and re-sanctioning of a single model as it moves from one domain to another. The migration of term structure models from academic finance into the derivatives markets cannot be adequately understood as the migration of a single model across domains, but instead must be examined as a longer-term translation process involving the development of dozens of new models that became gradually better aligned with the modelling culture practiced by derivatives practitioners. In this respect, our account shares similarities with Knuuttila and Loettgers's (2014) discussion of the migration of the Ising model from physics into the study of neural networks, which took the form of what is known as the Hopfield model (Knuuttila & Loettgers, 2014: 290). As they explain, the Hopfield model did not represent a direct analogical application of the Ising model to a new domain, but selectively adopted components of the model and added others appropriate to the new domain.

Finally, this case highlights the importance of model translation as both an enabler and constraint on model performativity. On the one hand, we see that modifications to term structure models—most notably, changes to allow them to "match" a given yield curve and the prices of vanilla interest rate options—were critical in allowing them to function as practical tools for pricing and hedging exotic derivatives. In making these changes, however, term structure models came to reflect, rather than challenge, the modelling culture of derivatives practitioners. In some respects, then, the case I present here is the inverse of that of the classic Black-Scholes-Merton case of performativity examined by MacKenzie and Millo (2003). There, a model shapes the market and the organization of derivatives trading. Here, a series of models are gradually translated to reflect the organization of the market and the prices produced by particular trading desks. As such, the case thus highlights the importance of translation in both allowing models to become embedded into the structure of markets but also in constraining their ability to shape markets. While translation is an implicit theme in MacKenzie's (2006) study of the performativity of the Black-Scholes-Merton model for pricing options, as D'Adderio et al. (2019) suggest, much of the controversy over Callon's concept of performativity implicitly assumes that models and theories are discrete entities that are able to migrate into new fields and organizations unchanged and exert one-way effects on the fields and organizations adopting them. For this reason, critics of the concept of the performativity of economics pointed to the need for establishing scope conditions on the performative power of models and theories, without which the performativity thesis implies that models and theories are simply "self-fulfilling prophecies" that become true only because they are accepted as such (c.f Ferraro & Pfeffer, 2005; Felin & Foss, 2009; Marti & Gond, 2018). Callon (2007), by contrast, instead emphasized that performativity implies that there will be a "long sequence of trial and error, reconfigurations and reformulations" of "articulating, experimenting, and observing" between a model and the world it describes. The case that I present in this

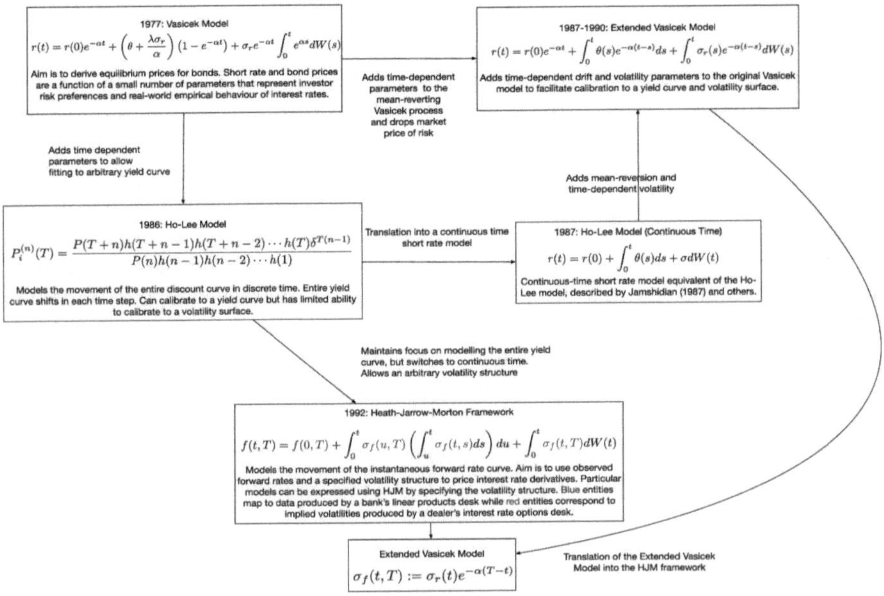

Fig. 8.3 Timeline of term structure model translation. Throughout the 1980s–1990s, models were gradually modified to embody an ontological separation between forward curves and volatility surfaces, a fundamental that reflects the modelling ontology of derivatives practitioners and mirrors the organizational division between trading desks

chapter suggests that the performative agency of models will depend on how they are translated into new domains. In this view, models are not all-powerful agents that are capable of re-shaping the social world on their own, as some critics of the performativity literature suggest that the concept implies (Fig. 8.3).

References

Baxter, M. W.. 1997. *General Interest-Rate Models and the Universality of HJM*. http://www.martinbaxter.co.uk/pubs/girm.pdf

Becker, G. S. (1990). *The economic approach to human behavior*. University of Chicago Press. http://books.google.com/books?hl=en&lr=&id=qQAZnc-mMSoC&pgis=1

Besedovsky, N. (2018). Financialization as calculative practice: The rise of structured finance and the cultural and calculative transformation of credit rating agencies. *Socio-Economic Review, 16*(1), 61–84. https://doi.org/10.1093/ser/mwx043

Beunza, D., & Ferraro, F. (2018). Performative work: Bridging performativity and institutional theory in the responsible investment field. *Organization Studies*, February, 0170840617747917. https://doi.org/10.1177/0170840617747917

Beunza, D., & Stark, D. (2004). Tools of the trade: The socio-Technology of Arbitrage in a wall street trading room. *Industrial and Corporate Change, 13*(2), 369–400. https://doi.org/10.1093/icc/dth015

Black, F., Derman, E., & Toy, W. (1990). A one-factor model of interest rates and its application to Treasury bond options. *Financial Analysts Journal, 46*(1), 33–39.

Bradley, S., & Thébault, K. P. Y. (2019). Models on the move: Migration and imperialism. *Studies in History and Philosophy of Science Part A, 77*(October), 81–92. https://doi.org/10.1016/j.shpsa.2017.11.008

Brennan, M. J., & Schwartz, E. S. (1979). A continuous time approach to the pricing of bonds. *Journal of Banking & Finance, 3*(2), 133–155.

Callon, M. (2007). What does it mean to say economics is performative? In D. MacKenzie, F. Muniesa, & L. Siu (Eds.), *Do economists make markets?: On the performativity of economics* (pp. 311–357). Princeton University Press.

Cassar, D. (2022). *Inscribing markets, shaping policy: A sociological investigation into the yield curve*. June. https://doi.org/10.7488/era/2415.

Cetina, K., Karin, T. R., Schatzki, E., von Savigny, K. D., Von Knorr-Cetina, E., Von Savigny, E., Savigny, T. R., Schatzki, K. K., & Cetina, and Eike Von Savigny. (2001). Objectual Practice. In *The practice turn in contemporary theory*. Taylor & Francis Group. http://ebookcentral.proquest.com/lib/ed/detail.action?docID=235322

Cheyette, O. (1992). *Markov representation of the Heath-Jarrow-Morton model*. BARRA Inc. http://papers.ssrn.com/sol3/papers.cfm?abstract_id=6073

Cheyette, O. (2002). Interest rate models. In F. J. Fabozzi (Ed.), *Advances in fixed income valuation: Modelling and risk management* (pp. 3–26). Wiley.

Chiapello, E., & Walter, C. (2016). The three ages of financial quantification: A conventionalist approach to the financiers, metrology. *Historical Social Research, 41*(2), 155–177. https://doi.org/10.12759/hsr.41.2016.2.155-177

Coombs, N., & van Der Heide, A. (2018). Financialization as Mathematization: The calculative and regulatory consequences of risk management. In P. Mader, D. Mertens, & N. van der Zwan (Eds.), *The Routledge international handbook of Financialization*. Routledge. https://osf.io/mptgs

Cox, J. C., Ingersoll, J. E., Jr., & Ross, S. A. (1985). A theory of the term structure of interest rates. *Econometrica, 53*(2), 385–407.

Cox, J. C., & Ross, A. (1976). The valuation of options for alternative stochastic processes. *Journal of Financial Economics, 3*(20292), 145–166.

Culbertson, J. M. (1957). The term structure of interest rates. *The Quarterly Journal of Economics, 71*(4), 485–517.

Czarniawska, B., & Sevon, G. (1996). *Translating organizational change*. Walter de Gruyter.

D'Adderio, L., Glaser, V., & Pollock, N. (2019). Performing theories, transforming organizations: A reply to Marti and Gond. *Academy of Management Review, 44*(3), 676–679. https://doi.org/10.5465/amr.2018.0378

Felin, T., & Foss, N. J. (2009). Social reality, the boundaries of self-fulfilling prophecy, and economics. *Organization Science, 20*(3), 654–668. https://doi.org/10.1287/orsc.1090.0431

Ferraro, F., & Pfeffer, J. (2005). Economics language and assumptions: How theories can become self-fulfilling. *The Academy of Management Review, 30*(1), 8–24.

Hagan, P. S., & West, G. (2006). Interpolation methods for curve construction. *Applied Mathematical Finance, 13*(2), 89–129. https://doi.org/10.1080/13504860500396032

Hagan, P. S., & Woodward, D. E. (1999). Markov Interest Rate Models. *Applied Mathematical Finance, 6*(4), 233–260. https://doi.org/10.1080/13504869950079275

Hansen, K. B. (2020). Model Talk: Calculative cultures in quantitative finance. *Science, Technology, & Human Values*, July, 0162243920944225. https://doi.org/10.1177/0162243920944225

Harrison, J., & Kreps, D. (1979). Martingales and arbitrage in multiperiod securities markets. *Journal of Economic Theory, 20*(3), 381–408. https://doi.org/10.1016/0022-0531(79)90043-7

Heath, D., Jarrow, R., & Morton, A. (1992). Bond pricing and the term structure of interest rates: A new methodology for contingent claims valuation. *Econometrica, 60*(1), 77–105.

Hicks, J. (1946). *Value and capital: An inquiry into some fundamental principles of economic theory, 1978*. Clarendon Press. http://books.google.com/books?id=1uwaz58_TN0C&pgis=1

Himick, D., & Brivot, M. (2018). Carriers of ideas in accounting standard-setting and financialization: The role of epistemic communities. *Accounting, Organizations and Society, 66*(April), 29–44. https://doi.org/10.1016/j.aos.2017.12.003

Hull, J., & White, A. (1994a). Numerical procedures for implementing term structure models I: Single-factor models. *Journal of Derivatives, 2*, 7–16.

Hull, J., & White, A. (1994b). Numerical procedures for implementing term structure models II: Two-factor models. *Journal of Derivatives, 2*, 37–48.

Humphreys, P. (2002). Computational Models. *Philosophy of Science, 69*(S3), S1–S11. https://doi.org/10.1086/341763

IFRS Foundation. (2013). *International financial reporting standard 13: Fair value measurement*.

Jamshidian, F. (1996). Bond and option evaluation in the Gaussian interest rate model. In L. Hughston (Ed.), *Vasicek and beyond: Approaches to building and applying interest rate models*. Risk Publications.

Jensen, C. B., & Morita, A.. (2016). Introduction: Infrastructures as ontological experiments. *Ethnos*, January. https://www.tandfonline.com/doi/abs/10.1080/00141844.2015.1107607

Kennedy, D. (2017). The machine in the market: Computers and the infrastructure of Price at the New York stock exchange, 1965–1975. *Social Studies of Science, 47*(6), 888–917. https://doi.org/10.1177/0306312717739367

Knorr-Cetina, K. (2000). *Epistemic cultures: How the sciences make knowledge*. Harvard University Press.

Knuuttila, T., & Loettgers, A. (2014). Magnets, spins, and neurons: The dissemination of model templates across disciplines. *The Monist, 97*(3), 280–300. https://doi.org/10.5840/monist201497319

Latour, B. (1984). The powers of association. *The Sociological Review, 32*(S1), 264–280. https://doi.org/10.1111/j.1467-954X.1984.tb00115.x

Lévi-Strauss, C. (1968). *The savage mind*. Second Impression edition. Weidenfeld and Nicolson.

MacKenzie, D. (2003). An equation and its worlds: Bricolage, exemplars, disunity and performativity in financial economics. *Social Studies of Science, 33*(6), 831–868.

MacKenzie, D. (2017). On invoking "culture" in the analysis of behavior in financial markets. In K. Chemla & E. F. Keller (Eds.), *Cultures without culturalism: The making of scientific knowledge*. Duke University Press.

MacKenzie, D. A. (2006). *An engine, not a camera: How financial models shape markets*. MIT Press.

MacKenzie, D., & Millo, Y. (2003). Constructing a market, performing theory: The historical sociology of a financial derivatives exchange. *American Journal of Sociology, 109*(1), 107–145. https://doi.org/10.1086/374404

MacKenzie, D., & Spears, T. (2014). "The formula that Killed Wall street": The Gaussian copula and modelling practices in investment banking. *Social Studies of Science, 44*(3), 393–417. https://doi.org/10.1177/0306312713517157

Mäki, U. (2009). Economics imperialism: Concept and constraints. *Philosophy of the Social Sciences, 39*(3), 351–380. https://doi.org/10.1177/0048393108319023

Marti, E., & Gond, J.-P. (2018). When do theories become self-fulfilling? Exploring the boundary conditions of performativity. *Academy of Management Review, 43*(3), 487–508. https://doi.org/10.5465/amr.2016.0071

Mikes, A. (2009). Risk management and calculative cultures. *Management Accounting Research, 20*, 18–40.

Minnis, M. A. 2013. *Correct numeraire/bond correlation in one-factor HJM models*. SSRN Scholarly Paper. Rochester, NY. https://doi.org/10.2139/ssrn.2308141

Mirowski, P., & Nik-Khah, E. (2008). Command performance: Exploring what STS thinks it takes to build a market. In T. Pinch & R. Swedberg (Eds.), *Living in a material world: Economic sociology meets science and technology studies* (pp. 89–128). MIT Press.

Modigliani, F., & Sutch, R. (1966). Innovations in interest rate policy. *The American Economic Review, 56*(1), 178–197.

Mol, A. M. (2003). *The body multiple*. Duke University Press. https://www.dukeupress.edu/the-body-multiple

Morgan, M. S. (2012). *The world in the model: How economists work and think*. Cambridge University Press.

Muniesa, F., Chabert, D., Ducrocq-Grondin, M., & Scott, S. V. (2011). Back-office intricacy: The description of financial objects in an investment Bank. *Industrial and Corporate Change, 20*(4), 1189–1213. https://doi.org/10.1093/icc/dtr020

Pinzur, D. (2021). Infrastructure, ontology and meaning: The endogenous development of economic ideas. *Social Studies of Science*, April, 03063127211011524. https://doi.org/10.1177/03063127211011524

Rebonato, R. (2002). *Modern pricing of interest-rate derivatives: The LIBOR market model and beyond*. Princeton University Press.

Rebonato, R. (2004a). Interest-rate term-structure pricing models: A review. *Proceedings of the Royal Society A: Mathematical, Physical and Engineering Sciences, 460*(2043), 667–728. https://doi.org/10.1098/rspa.2003.1255

Rebonato, R. (2004b). *Volatility and correlation: The perfect hedger and the fox*. Wiley.

Riles, A. (2010). Collateral expertise. *Current Anthropology, 51*(6), 795–818. https://doi.org/10.1086/657627

Roll, R. (1970). The behavior of interest rates: An application of the efficient market model to U.S. *Treasury Bills*. Basic Books. http://books.google.com/books?id=bbGxAAAAIAAJ&pgis=1

Spears, T. (2014). *Engineering value, engineering risk: What derivatives quants know and what their models do*. University of Edinburgh.

Svetlova, E. (2014). Modelling beyond application: Epistemic and non-epistemic values in modern science. *International Studies in the Philosophy of Science, 28*(February 2015), 79–98. https://doi.org/10.1080/02698595.2014.915656

Svetlova, E. (2018). *Financial models and society: Villains or scapegoats?* Edward Elgar Publishing.

Svetlova, E., & Dirksen, V. (2014). Models at work—Models in decision making. *Science in Context, 27*(04), 561–577. https://doi.org/10.1017/S0269889714000209

Tuckman, B. (2002). Fixed Income Securities: Tools for Today's Markets. John Wiley & Sons.

van Der Heide, A. (2019). Model migration and rough edges: British actuaries and the ontologies of modelling. *Social Studies of Science*, December, 0306312719893465. https://doi.org/10.1177/0306312719893465

Vašícek, O. (1977). An equilibrium characterization of the term structure. *Journal of Financial Economics, 5*(2), 177–188. https://doi.org/10.1016/0304-405X(77)90016-2

Wilmott, P. (2000). *Paul Wilmott on quantitative finance*. Wiley. http://books.google.com/books?id=xKpOlK2oHKwC&pgis=1

Zaloom, C. (2009). How to read the future: The yield curve, affect, and financial prediction. *Public Culture, 21*(2), 245. https://doi.org/10.1215/08992363-2008-028

Chapter 9
Advanced Trading Systems: Aggressive and Unfair Speculation or Instruments Featuring a Positive Impact on Overall Markets Quality? A Comprehensive Assessment

G. Cellai and A. Puorro

Abstract In this chapter the authors offer their perspective about modern trading systems, widely known as algorithmic traders (ATs) or high frequency traders (HFTs). At a glance, their presence might be considered unfair and unhealthy: their activity is often perceived as purely speculative, based on the edge that such systems have over other market participants, often less informed and slow. However, ATs and HFTs are a natural consequence of technological progress and they are a relevant and unavoidable component of modern financial ecosystems. Therefore, a holistic and comprehensive assessment of their impact on overall market quality seems useful for a better understanding of their actual role. As in the case of many other phenomena that influence our daily life in a pervasive way, a deeper insight is often more valuable than a moral judgment, as it helps building up adaptive strategies and live with them.

9.1 Introduction

Since the early 2000s, financial markets have experienced a wide range of structural changes. Increased processing power, reduced storage costs, and the ability to access, analyze, and transfer vast amounts of data almost instantaneously have led to a robust, and in many cases dominant, presence of non-human trading activity.

The views expressed in the articles are those of the authors and do not involve the responsibility of the Bank.

G. Cellai · A. Puorro (✉)
Bank of Italy, Rome, Italy
e-mail: gioia.cellai@bancaditalia.it; alfonso.puorro@bancaditalia.it

E. Ippoliti et al. (eds.), *Philosophy and Finance: Ten Open Questions*, Synthese Library 504, https://doi.org/10.1007/978-3-031-96510-4_9

Algorithmic trading and high-speed speculation represent a significant evolution in financial markets, where advanced algorithms and high-frequency trading (HFT) technologies are used to execute trades at incredible speeds.

The essence of algorithmic trading lies in the automation of the decision-making process, where algorithms—sets of complex mathematical instructions—analyze vast amounts of market data, identify trading opportunities, and execute orders, all within fractions of seconds. These algorithms are designed to respond to market conditions in real-time, leveraging upon high-speed data processing and analysis to capitalize even on the smallest price fluctuations. Recently, cutting-edge Artificial Intelligence (AI) technologies, particularly Machine Learning (ML), have taken center stage, further enhancing these processes. The application of ML techniques in trading has expanded significantly. For example, ML is now used to analyze a wide range of macroeconomic data for optimizing timing in cross-asset investments, to employ deep learning approaches for tasks such as portfolio management, and to analyze unstructured data sets, including social media feeds and satellite imagery, so as to support predictive capacity of economic and financial variables.

Modern speculation specifically targets price discrepancies that exist for only fractions of a second, aiming to generate profit through rapid trades. High-speed speculation relies heavily on the principle of arbitrage, namely the profits arising from price differences of the same or similar financial instruments across different markets or platforms. These opportunities are short-lived, often disappearing within milliseconds, which is why the speed of execution is paramount. The algorithms must not only detect these price discrepancies in real-time but also execute the corresponding trades instantaneously to capitalize on them before the market adjusts.

The efficiency and profitability of high-speed speculation continue to drive its adoption in financial markets. For trading firms, the ability to execute trades at lightning speeds represents a significant source of revenue.

Such changes have enabled participants to respond to developments in near-real time and have paved the way for, significant efficiency gains, and new business models.

The extent of automated trading activity varies across asset classes. Official estimates suggest ranges from 30% to 70%, indicating that a significant portion of trading is driven by non-human decisions. According to Chaboud et al. (2023), the use of *manual trading* in the global FX market has been steadily declining since 2024, reaching a multiyear low of 13% in 2022. This decline has been accompanied by a constant increase in non-human trading activity, which has now become the dominant trading method. Alongside these technological advancements, financial markets have seen the rise of new types of participants, in particular non-bank Proprietary Trading Firms (PTFs). These firms trade on a proprietary basis, often at high frequencies and using sophisticated technology, and can act as both liquidity providers and consumers. Specifically, for PTFs alone, their size is often challenging to quantify, but they are estimated to account for approximately 50% of trading in both US equities and on-the-run US Treasuries.

Against this background, according to the BIS (BIS/Markets Committee Report, 2018),[1] modern financial markets can be described as *fast-paced*, meaning that the price discovery process predominantly occurs through electronic means and is characterized by a substantial presence of high-speed, algorithmic activity.

As a result, the ecosystem of modern financial markets drastically changed. Algorithmic trading and high-speed speculation are at the forefront of modern financial markets, leveraging on cutting-edge technology to execute trades with unparalleled speed and precision. While these practices have the potential to enhance market efficiency and liquidity, they also introduce challenges and risks, particularly concerning market welfare. Profitability has become increasingly difficult; new entrants have emerged, and established players have adapted and evolved. Automation has become essential. In a zero-sum game, gains achieved through technological advancements must be offset by less sophisticated participants.

Such picture may raise critical questions on the fairness of AT and HFT and on the 'legitimacy' of their presence in the market, given the huge gap between human traders and sophisticated algorithms that leverage upon immense speed and precision. This paper does not aim at addressing such questions nor at providing a moral or ethical judgement on this new generation of market players. Based on the view that *a deeper understanding of complex phenomena is the cornerstone of adaptive strategies that are necessary to survive in a rapidly changing environment*, the following sections try to provide a thorough and unbiased analysis.

The first section goes through the main fields of action covered by AT and HFTs. In the second part, implications of the presence of such players in financial markets are examined in a balanced and comprehensive way. Finally, some conclusions are drawn, with a focus on how such players tend to affect market dynamics in different states of the world.

9.2 The Main Fields of Action of Modern Trading Systems

9.2.1 High Frequency Traders as Modern Arbitrageurs

Arbitrage activity entails systematically analyzing financial markets to detect price discrepancies of identical or highly correlated securities across different trading venues. This strategy is not limited to perfectly identical assets but can also involve instruments that are considered close substitutes. The goal is to exploit such mispricings to generate risk-free or low-risk profits.

Starting from the early 2000s the trading industry has been investing in new technologies designed to minimize the time needed to detect and execute arbitrage opportunities (Lewis, 2015). In a few years such developments reached the physically possible limits, so that nowadays arbitrage opportunities last the blink of an

[1] Report submitted by a Study Group established by the Markets Committee. September 2018.

eye,[2] hardly pushing competition across players. Contemporary arbitrage strategies differ markedly from traditional ones, as they can only be leveraged by a select group of traders equipped with cutting-edge technology and extensive expertise. As a result, these strategies are no longer viewed as mere arbitrage but are redefined as *latency arbitrage*, reflecting their dependence on minimizing latency to capitalize on fleeting market inefficiencies

High Frequency Traders (HFTs) are the only players able to exploit latency arbitrages. They entered the U.S. equity market in 2004, at a time when it had become highly fragmented due to the emergence of parallel trading venues known as Electronic Communication Networks (ECNs[3]) and regulatory enhancements aimed at fostering fair competition.[4] In response, traders were compelled to develop technology to boost computational power and reduce reaction times. Recognizing that profitability increasingly depended on speed, some traditional traders explored sophisticated ways to establish a competitive edge over other market participants.

HFT is a form of algorithmic trading that utilizes powerful computational capacities to execute a vast number of trades in fractions of a second. The primary goal of HFT is to capitalize on tiny price discrepancies that exist for only milliseconds, generating profit through rapid and frequent trades. This trading strategy relies heavily on sophisticated algorithms and massive computing power to analyze market data, identify trading opportunities, and execute orders faster than human traders or traditional automated systems could possibly achieve. The computational capacity required for HFT is enormous, involving the processing of vast quantities of data in real-time. This includes not only the current prices of securities across different exchanges but also other data points, such as order book information. The algorithms used in HFT must be able to deconstruct this data quickly, make complex decisions based on pre-programmed strategies, and execute trades almost instantaneously.

Latency is a physical concept more than a financial one; it represents the amount of time necessary for an electrical impulse (the trading decision) to move from place A (the initiating place) to place B (trading venue's servers). In financial words, latency measures the (fractions of) seconds necessary to turn a trading decision into an effective negotiation. It denotes the minimal delay between the reception

[2] What was meant by "fast" was changing rapidly. In the old days—before, say, 2007—the speed with which a trader could execute had human limits. Human beings worked on the floors of the exchanges, and if you wanted to buy or sell anything you had to pass through them. The exchanges, by 2007, were simply stacks of computers in data centers. The speed with which trades occurred on them was no longer constrained by people. The only constraint was how fast an electronic signal could travel between Chicago and New York—or, more precisely, between the data center in Chicago that housed the Chicago Mercantile Exchange and a data center beside the Nasdaq's stock exchange in Carteret, New Jersey.

[3] Electronic Communications Networks, or ECNs, as defined in Rule 600(b) of SEC Regulation National Markets Securities, are electronic trading systems that automatically match buy and sell orders at specified prices.

[4] Regulation National Market System (Regulation NMS), 2005.

of market data and the execution of a trade. It is a critical metric that reflects the efficiency of trading algorithms in processing information and executing orders. In a highly competitive environment, the speed at which such decisions are made is crucial, as the competitive advantage in high-speed speculation often hinges on being faster than other market participants. To achieve this, trading firms invest in cutting-edge technologies, such as low-latency networks, co-location services (where traders place their servers close to exchange servers to minimize data transmission time), which allow for ultra-fast data processing and execution of trades.

Latency is a necessary but insufficient condition for achieving profitability. The second essential factor is *computational capacity*, which is crucial for monitoring the financial landscape and converting informational input into actionable trading decisions. Enhanced computational capacity markedly improves the identification of arbitrage opportunities within financial markets. Sophisticated algorithms and high-performance computing systems facilitate the rapid analysis of extensive market data, allowing traders to detect pricing discrepancies across various venues. With advanced computational resources, arbitrageurs can capitalize on transient opportunities before they vanish.

9.2.2 HFT and Speculation

Initially, speed and computational capacity were primarily developed to lock in arbitrage-based profits. However, as market conditions evolved, the profitability of pure arbitrage opportunities waned. Therefore, the investment community soon wondered it advanced and ultra-fast automated trading strategies could also be used for speculative activities. Consequently, in recent years, HFT firms have increasingly transitioned towards speculative trading.

HFT technology rapidly proved incredibly useful in order to extract profit. Sophisticated algorithms were soon programmed so as to analyze vast amounts of data, to predict and react to market trends. Central to HFT's success are various speculative strategies designed to benefit from fleeting opportunities. Each strategy aims to achieve profitability through rapid execution and real-time analysis of market data. Understanding these strategies provides insight into the complex dynamics of contemporary financial markets and the role of technology in shaping trading behaviors.

- *Momentum trading—Trend Following*. This strategy identifies and capitalizes on short-term price trends. HFT firms employ established statistical models (Lo et al., 2000) to detect and follow these trends, aiming to profit from brief periods of sustained price movements. Momentum trading algorithms continuously analyze market, looking for patterns such as sudden price movements or high trading volumes that indicate strong momentum (Menkveld & Yueshen, 2019). Once a trend is identified, the algorithm executes numerous trades to align with the prevailing momentum. These trades are executed at extremely high speeds, allowing traders to enter and exit positions rapidly in response to minor price fluctuations. Momentum trading algorithms also utilize sophisticated techniques to manage positions and maximize profits, including adjusting trade sizes and exit points based on the trend's strength and current market conditions.

- *Liquidity Detection—Stop Haunting.* Liquidity detection involves identifying large orders or liquidity pools hidden in the market. HFT systems analyze order flow to anticipate and react to significant trades or liquidity changes, positioning themselves to benefit from price movements (Menkveld, 2013). Liquidity detection algorithms continuously monitor real-time market data, including order book information, trading volumes, and bid-ask spreads. They use these data to identify patterns that indicate the presence of large or hidden orders. The strategy involves analyzing the order book to detect large limit orders that may not be immediately visible to other market participants. These large orders, known as "liquidity pools," can provide insights into potential price movements or significant trading activity. By detecting large orders or hidden liquidity, algorithms can anticipate price movements and adjust their trading strategies accordingly. For example, if a large buy order is detected, the algorithm might place buy orders in advance to benefit from the anticipated increase in price. A particular liquidity detection strategy is called *stop haunting*, designed to exploit the stop-loss orders placed by other market participants. Stop-loss orders are used by traders to limit potential losses by automatically selling a security when its price falls to a certain level. The Stop Haunting strategy involves monitoring these stop-loss levels and executing trades to trigger them. Traders using stop haunting strategy place their own orders strategically to cause the market price to reach or breach the stop-loss levels set by other traders. By pushing the price to these levels, they can force the execution of stop-loss orders, which may lead to additional selling pressure and further price declines. This creates a cascade effect where the trader benefits from the subsequent price movements that result from the triggered stop-loss orders. The primary goal of Stop Haunting is to profit from the volatility induced by the execution of stop-loss orders. Traders using this strategy aim to capture profits from the resulting price movements, which are often amplified by the collective reaction of many stop-loss orders being triggered. Part of high-speed speculation in the realm of liquidity detection strategies focuses on exploiting opportunities that arise and vanish in milliseconds. *Flash trading* algorithms leverage high-speed data feeds and advanced analytics to continuously monitor order book data, price movements, and other market indicators to identify potential trading opportunities. Their technology, in particular based on ML techniques, allows them to front-run incipient, but still not concrete, directional trading flows, taking advantage of them. Flash traders exploit delays in the transmission of market data between different exchanges and different categories of market operators, as a consequence, such a strategy relies on executing trades at extremely high speeds (within milliseconds), requiring very low latency. By acting on information before it is fully disseminated, they can collect non negligible gains.

- *Quote Stuffing* involves submitting a large volume of orders to overwhelm market systems and create temporary price movements (CFTC and SEC, 2010[5]). This strategy is often used to create opportunities for other HFT strategies, although it can be seen as manipulative and is subject to regulatory scrutiny.

9.2.3 Market Making and Liquidity Providing

Along with pure speculative strategies, modern trading systems also implement strategies that are closely related to the essential services provided by market makers for maintaining a healthy market ecosystem. Nonetheless, these strategies are increasingly being pursued in a more speculative manner, blurring the lines between traditional market-making functions and speculative trading practices.

Market makers consistently provide liquidity by offering to buy and sell particular securities or assets. They post bid (buy) and ask (sell) prices for these securities, ready to execute trades with other market participants. The difference between the bid and ask prices, known as the bid-ask spread, is where market makers generate their profit. This spread compensates them for the risks they take on by holding an inventory of the security and potentially trading with more informed participants. Market makers play a crucial role in ensuring that markets remain liquid and efficient. By continuously providing quotes, they make it easier for other traders to enter and exit positions, reducing the time it takes to find a counterparty and minimizing market risk. In fast-moving or less liquid markets, the presence of market makers helps prevent large price swings by absorbing buy and sell orders, thus stabilizing the market.

High-frequency technology has been rapidly adopted to manage market-making risks. Market making involves considerable risks, particularly during periods of high volatility or when there is a sudden inflow of unidirectional informed trades who may take advantage of the market maker's quotes. Therefore, they started to use sophisticated algorithms based on high frequency technology to adjust their quotes and hedge their positions, reducing the impact of sudden and significant price swings.

Market makers' profits are attractive, leading the trading community to develop strategies that mimic their activities, while avoiding the associated risks and duties. In other words, HFTs are increasingly substituting the activity of market makers to exploit the benefits of market making without assuming the associated obligations.

- *Liquidity providing—pseudo Market Making*. It is a strategy where HFTs continuously offer to buy or sell securities, thereby supplying liquidity to the market. This strategy is fundamental to market-making activities, where HFT firms aim to profit from the bid-ask spread. Liquidity providers place numerous buy and

[5] CFTC & SEC. (2010). *Findings Regarding the Market Events of May 6, 2010*. U.S. Commodity Futures Trading Commission and Securities and Exchange Commission.

sell orders for a security at various price levels. These orders are continuously updated based on market conditions, ensuring that the trader is always ready to transact at both ends of the bid-ask spread. Liquidity providers actively manage their order book to ensure they are offering competitive prices. At the same time, they monitor market conditions, including order flows and volatility, to adjust their bid-ask spreads and optimize profitability. HFT firms rely on advanced technology to update their quotes and execute trades at extremely high speeds. This allows them to quickly respond to market changes and minimize the risk associated with holding positions for too long (Biais et al., 2015).

9.3 The Impact of Modern Trading

9.3.1 Modern Trading Strategies and Market Functioning

The use of algorithms in high-speed speculation is extremely contentious, generating debate regarding its implications for market quality. Some argue that algorithms enhance market efficiency by improving overall liquidity, which benefits all participants by lowering transaction costs and facilitating price discovery. However, the rapid and automated nature of algorithmic trading can exacerbate market volatility and amplify systemic risks. Concerns arise in particular about fairness, as sophisticated algorithms may give certain traders an undue advantage over others, potentially leading to unequal access and manipulation risks. For the sake of simplicity, the considerations below are mostly referred to HFT, but they can be broadly applied to the entire spectrum of modern, advanced trading systems.

A robust strain of academic research tried to assess the impact of HFT shortly after its widespread adoption, especially in the aftermath of market crashes for which HFTs were blamed by the financial press and market practitioners. However, their positive contribution to overall market quality is also widely acknowledged.

Market quality is strictly related to market functioning; it pertains to the overall effectiveness and efficiency featuring the trading process, encompassing several key dimensions, with particular regard to liquidity, price discovery and volatility.

- *Liquidity* is a fundamental component of market quality. It indicates how easily assets can be bought or sold without causing significant price changes. High liquidity is typically associated with narrower bid-ask spreads and lower transaction costs, which enhance the overall attractiveness of the market (Chordia et al., 2001). Liquidity should be valued in a multifaceted manner, including more sophisticated metrics, such as robustness (the capacity of a market to absorb large order flow imbalances with limited price swings) and resilience (The capacity of a market to reach fair values after experiencing significant order flow imbalances). Liquidity is closely linked to *Transparency, i.e.* the availability and clarity of information regarding market activities and trading processes. Liquid and transparent markets allow participants to make informed decisions based on

readily accessible and reliable data (Bessembinder, 2003). High market quality is achieved when there is a high level of transparency, reducing the likelihood of manipulation and fostering trust among market participants.
- *Price discovery* refers to the process through which market prices reflect all available information. Efficient price discovery ensures that securities are priced accurately according to their intrinsic value and current market conditions (Grossman & Stiglitz, 1980). A market with high-quality price discovery provides more accurate signals to investors and reduces informational asymmetries.
- *Volatility* measures the degree of variation in asset prices over time. While some volatility is physiological in financial markets, excessive volatility can indicate poor market quality and lead to investor uncertainty and instability (Black, 1976).

The presence of HFT is indeed generally viewed as a beneficial force in improving all the above metrics of market quality. Numerous researchers contend that HFT can exert a positive impact on market liquidity by effectively tightening bid/offer spreads and reducing transaction costs (Hendershott et al., 2011). The liquidity providing activity pursued by most HFTs supports also more refined metrics of liquidity, such as trading books robustness and resilience. This process makes it easier for buyers and sellers to transact at more favorable prices, allowing market participants to make more informed trading decisions and better align their strategies with real-time market conditions (Angel et al., 2010; Brogaard, 2010).

Through its algorithmic precision and speed, it provides a continuous stream of market trades, enhancing the efficiency of price discovery processes. It is a fact that fast momentum strategies allow for an almost instantaneous alignment of prices with available information, positively contributing to a faster and sounder price discovery process (Angel et al., 2010) as well as market efficiency.

HFTs' activity seems also beneficial with regard to market volatility. Although HFTs can contribute to an increase in volatility in the immediate aftermath of the release of economic news, such increase can be viewed as an intense and rapid price discovery process, which supports a swift convergence of prices toward fair levels. This, in turn, may help mitigate further volatility and enhance overall market stability. Moreover, the swift execution of trades reduces price discrepancies, contributing to more efficient markets. Consequently, HFT plays a crucial role in dampening excessive fluctuations and promoting smoother price movements (Brogaard et al., 2014; Hasbrouck & Saar, 2013).

While HFT's role in enhancing market liquidity and efficiency is well-documented, its overall impact remains a crucial area of ongoing research and debate. The evolution of HFT from arbitrageurs to speculators and pseudo market makers has sparked concerns about market stability, as speculative trading can exacerbate volatility and lead to more pronounced price swings. Critics argue that the speculative nature of modern HFT may contribute to market distortions and raise questions about the broader impact of such trading practices on financial markets fairness. Market volatility might be affected as well, as the rapid execution of large volumes of trades can exacerbate price swings, particularly in times of market stress. Specialized press often suggests that the most aggressive practices applied by HFTs may even generate situations encompassing a high probability of market inefficiency, aimed at increasing their profit opportunities at the expenses of other players.

The flash crash of 2010, when the Dow Jones Industrial Average plunged nearly 1000 points in a matter of minutes before quickly recovering, is often cited as an example of the potential risks associated with high-speed algorithmic trading: high-frequency algorithms, reacting to unusual market conditions contributed to create a feedback loop that amplified the sell-off, leading to a brief but dramatic market drop. Similar conclusions were reached with regard to another major episode happened on the US Treasury market in October 2014, when the US 10Y Treasury yield dropped by about 30 basis points in 30 min. The move was driven by a combination of factors, including the increasing role of HFT and the occurrence of large, sudden and unidirectional trades, in most cases automated. Such elements created a liquidity vacuum, amplifying the market's reaction to shifts in supply and demand related to the release of a macroeconomic data (CFTC and SEC Report, 2014).

In such context, information asymmetries, adverse selection, liquidity evaporation and flash crash phenomena are often mentioned as the worst effects of high frequency trading.

- *Information asymmetries* in financial markets emerge when traders have disparate levels of market knowledge, creating imbalances in the information available to different participants. This disparity introduces the risk that a less informed trader may inadvertently provide liquidity to more informed traders, leading to suboptimal trading outcomes. For instance, if a trader submits a limit order,[6] its view about market dynamics becomes visible to other participants. Under certain circumstances, a better-informed trader might exploit this position, potentially generating losses for the less-informed one. HFT algorithms can engage in "order anticipation strategies," predicting the movements of the limit order and executing trades that capitalize on its anticipated impact. In general, the potential of machine learning techniques, such as those utilizing satellite imagery, is crucial for acquiring and leveraging superior information. More ambiguous implications arise in flash trading strategies, where information asymmetries can come to the fore as undue advantages that might also extend beyond front-running practices[7] (Xu & Xue, 2021).
- *Adverse selection* is closely linked to information asymmetries, referring to a situation where investors make suboptimal investment choices due to low quality, misleading if not fake information. In the context of high-frequency trading strategies, adverse selection becomes particularly pronounced with the use of

[6] A limit order (LMT) is an instruction to buy or sell a security at a specified price or better, ensuring price control but not execution certainty.

[7] A flash trading strategy involves using HFT algorithms to detect and exploit short-lived order imbalances, executing trades in milliseconds before the market reacts. Those strategies capitalize of the capacity of HFTs not only to process, but also to access information before other market participants are able to obtain it, therefore raising questions about their fairness.

stop hunting tactics. This strategy exploits the information asymmetry between traders who place stop-loss orders and those who anticipate their execution. By pushing the market to reach these stop-loss levels, traders can benefit from the forced selling or buying, creating further inefficiencies. This practice not only raises significant concerns about fairness and market integrity but can also contribute to increased market instability by amplifying price distortions and volatility (Foucault & Moinas, 2018).

- *Liquidity evaporation* is a byproduct of liquidity providing strategies adopting HFT technology. It occurs when the liquidity displayed on order books vanishes almost instantaneously, in response to sudden information release or episodes of volatility. HFT firms, which provide a significant portion of market liquidity, are not subject to traditional market makers' obligations in terms of stability and reliability of the liquidity provided. They may quickly withdraw or adjust their orders to avoid unidirectional order flow threatening their profitability. HFT firms operate on razor-thin margins, their algorithms are designed to minimize risk exposure by quickly cancelling orders if they sense potential losses or instability. This behavior can create a feedback loop, where the reduction in liquidity leads to greater price volatility, prompting further withdrawal of liquidity, thus posing risks to market stability, as it can exacerbate price swings and reduce the ability of other participants to execute trades efficiently (Degryse et al., 2020).
- *Volatility transmission.* Empirical results (Panzarino et al., 2016) highlight risks that—in times of market turmoil—the presence of HFT, although fostering price discovery, also enhances the transmission of volatility across substitute market segments. For example, an increase in HFTs activity on the BTP futures market may favor the transmission of volatility spikes to the spot segment, with negative consequences on the liquidity conditions of Italian Government bond secondary market.
- *Flash crashes.* Liquidity evaporation and adverse selection can contribute to extreme price swings and have been implicated in severe market events such as *flash crashes*, where rapid liquidity withdrawal, along with unidirectional order flows, lead to huge adverse selection phenomena, triggering stop losses ad further intensifying the prevailing market trend. Under certain circumstances the market can become completely dysfunctional, resulting in price swings by more than $\pm 10\%$ in a matter of minutes. Flash crashes should not be dismissed as minor events merely representing a redistribution of wealth among different investor categories. In some cases, they can evolve into systemic disruptions that also affect asset classes significant from an institutional perspective, such as those relevant to the conduct of monetary policy. From this viewpoint, they can have relevant public consequences, potentially leading to poor decision-making as well as the destruction of substantial public wealth (for example, if a crash occurs ahead of a public debt security auction, see Flora & Renò, 2020).

Without downplaying such undesired effects, an emerging stream of academic research adopts a more balances approach while assessing the impacts of HFT on market quality. Their nuanced findings suggest that the overall effect of HFT on market dynamics is more complex and context-dependent than previously understood. Against this background, such multifaceted approach differentiates between passive and aggressive algos within the HFT industry (Benos & Sagade, 2012; Kirilenko et al., 2017).

Aggressive HFTs are high-frequency traders who consume liquidity in the market rather than providing it. These traders tend to push the market in specific directions and act to follow or anticipate price trends rather than contributing to liquidity improvement. They place orders that remove liquidity from the market, such as buy orders that are executed immediately without waiting for the price to reach the desired level. They use strategies based on short-term price trend analysis and rapid market movements. Unlike *passive HFTs*, which provide liquidity and tend to maintain a neutral trading direction, aggressive HFTs are more likely to quickly change their positions in response to market movements and trading signals. Therefore, aggressive HFTs are perceived as more problematic, because—according to their intrinsic speculative nature—they are prone to absorb profits, at the expenses of other market participants.

While the distinction between passive and aggressive HFT provides a useful framework for analyzing potential impacts on market dynamics, the overall assessment of HFT's influence on market quality still remains complex. Passive HFTs, perceived as beneficial due to their continuous liquidity provision, are not without dark sides. In certain circumstances, their presence foster market fragility. Their algorithmic strategies might be structured in order to suddenly withdraw liquidity during periods of heightened, unexpected volatility, exacerbating price swings and leading to potential flash crashes (Kirilenko et al., 2017). Their liquidity is indeed considered transient, disappearing when most needed (Easley et al., 2012).

Conversely, aggressive HFTs, characterized by their liquidity-consuming strategies, often lead passive HFTs information contribution.[8] Aggressive HFTs exhibit greater informational efficiency, reacting more swiftly to new information and thus contributing significantly to price discovery (Brogaard et al., 2014). However, such efficiency comes at a cost, as their trading activities can also introduce noise, increasing short-term volatility and potentially destabilizing markets during periods of stress (Jovanovic & Menkveld, 2016).

While assessing high-frequency trading (HFT), differentiating between aggressive and passive trading strategies provides valuable insights; however, the assessment would be s incomplete without considering the broader context of market conditions.

The impact of HFT seems indeed inherently contingent on the prevailing state of the markets, which means that its effects are highly state-dependent. In stable market conditions, the presence of high frequency traders has mostly positive effects:

[8] Aggressive HFTs identify and react to emerging market conditions or exploit temporary inefficiencies. By executing trades swiftly, they influence price discovery and market movements. Consequently, passive HFTs, while contributing to market depth and stability, may find themselves responding to the price signals and market trends established by their aggressive counterparts, who dominate short-term market dynamics with their proactive trading strategies.

passive HFTs, which engage in strategies such as liquidity provision and market making, generally contribute to narrower bid-ask spreads and increased market efficiency. However, this favorable assessment of HFTs can change dramatically during periods of heightened volatility or market uncertainty. In such riskier environments, the behavior of HFTs, with particular regard to the aggressive ones, comes to the fore, suddenly and massively. Aggressive HFTs, which engage in high-speed trading to capitalize on short-term price movements, dominate market places, contributing to exacerbated price swings, increased market instability and reduced market welfare.

9.3.2 Fostering Greater Competition

All the strategies described above are aimed at generating profit, pushing modern traders in a position to understand price developments before other market participants and to act faster. By using HFT techniques, the ability to anticipate market movements enables them to capitalize on short-lived opportunities and capture small price discrepancies for profit. While the competitive advantage of modern traders over traditional ones may seem disproportionate, it's important to note that the competition among advanced traders is exceptionally intense.

In the realm of HFT, firms and traders are continually refining their algorithms and technologies to stay ahead. They invest heavily in infrastructure to maintain their competitive edge. This relentless pursuit of speed and efficiency creates a highly competitive environment where incremental improvements in technology or strategy can result in significant financial gains. As a result, the race among developed traders is not just about being faster but also about innovating and adapting to ever-evolving market conditions.

As shown in Fig. 9.1, in order to achieve significant profits (joining the group depicted in blue), an HFT firm must heavily invest in advanced technology, particularly in reducing *latency* and enhancing *computational power*. Latency requires substantial investments in high-speed data feeds and co-location services. Additionally, significant computational capacity is necessary to process large volumes of data and execute complex trading strategies rapidly, ensuring that the firm can remain competitive in a fast-evolving landscape. Profitable algorithms are often crafted by highly specialized teams including quantitative analysts, data scientists, and programmers. Such professionals possess rare and costly sets of skills, including advanced knowledge of mathematics, statistics, machine learning, and financial markets. The scarcity of similar talents, coupled with the high salaries and resources required to attract and retain them, makes the creation of profitable HFT algorithms both an expensive and exclusive endeavor, accessible only to well-funded firms (Fig. 9.1).

Moreover, these algorithms are often based on advanced mathematical models and require extensive back-testing against historical data to ensure their effectiveness. This back-testing process can be computationally intensive, as it involves

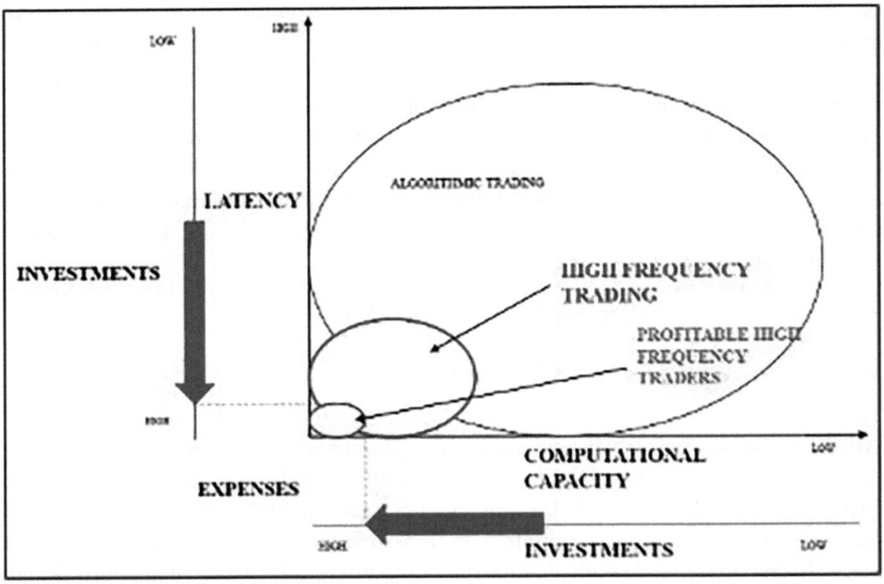

Fig. 9.1 Algorithmic trading

simulating the algorithm's performance over vast amounts of data to fine-tune its parameters and improve its predictive accuracy. Once deployed, the algorithm must continuously adapt to changing market conditions, which requires ongoing analysis and adjustment, again demanding substantial computational power. Markets are dynamic, and the strategies that work today might not work tomorrow; new data sources become available over time and HFT firms must adapt their strategies to incorporate such inputs. Success depends on staying ahead through continuous innovation and investment, as firms strive to outperform competitors by leveraging upon the latest tools and expertise to gain a competitive edge in the fast-evolving financial landscape.

9.3.3 The Effects on the Business Model and Transaction Fees of Trading Venues

HFT has significantly impacted the revenues of trading venues, influencing their investment strategies and cost structures. Their presence has implications for trading venues' revenue models, particularly in terms of commissions and fees generated from the high volume of trades they execute.

Trading venues earn revenue primarily through transaction fees, listing fees, data sales, and market access fees. HFT firms, which often engage in thousands or millions of trades per day, contribute significantly to the overall trading volume on exchanges, as the volume of their transactions is immense.

If we consider the New York Stock Exchange (NYSE) and Nasdaq, two of the largest U.S. trading venues, a study by Oxera (2011[9]) found that a significant portion of their revenues comes from transaction fees, especially those associated with high-frequency trading activity.[10] The presence of HFT firms, therefore, directly influences these revenues through their continuous trading activities.

While HFT increases transaction revenues for trading venues, it also intensifies competition among exchanges to attract HFT activity. HFT firms typically operate on very tight margins and are highly sensitive to trading costs. Such sensitivity induced several trading venues to adopt fee structures designed to attract HFT activity.[11] In order to gain HFT business, exchanges often lower fees, provide rebates, or make substantial investments in low-latency infrastructure.

As a matter of fact, trading venues rely on HFTs for increased trading volumes and revenue. At the same time, the presence of HFTs fosters competition among venues, with benefits for all typologies of customers in terms of lower fees. Additionally, the pressure to attract HFTs pushes exchanges to invest in advanced technology, leading to a more robust and efficient trading infrastructure.

9.4 Conclusion

The multifaceted impacts of advanced trading systems reflect the complexity of their role in financial markets.

HFTs' and modern trading systems' contribution to market quality is not uniform, as it changes in response to market conditions: while they enhance market efficiency and contribute to stability in quiet periods, in times of turmoil they tend to amplify market stress. On average, they contribute positively to market dynamics, often acting as a form of invisible white noise that provides liquidity and enhances market quality, extracting profits that would otherwise remain concealed. From a Pareto efficiency perspective, such role aligns with the principle of improving market efficiency without making anyone worse off. However, it is important to notice that HFTs can rapidly alter their behavior, transitioning from liquidity providers to liquidity consumers and voracious profit-seekers. During these, short

---

[9] Oxera paper, commissioned as part of the UK Government's Foresight Project on The Future of Computer Trading in Financial Markets. (2011). *"What is the economic impact of the MiFID rules aimed at regulating high-frequency trading?"*

[10] In 2020, Nasdaq reported approximately $2.9 billion in net revenues, with over 40% attributed to market services, which include trading revenues from both equities and derivatives.

[11] One popular model is the maker-taker pricing model, in which traders who provide liquidity (makers) receive a rebate, while those who take liquidity (takers) pay a fee. This model incentivizes HFT firms to place limit orders that add liquidity to the market, generating significant trading volume and, in turn, revenue for the venue.

lived but intense, phases they are undoubtedly a threat for other players, potentially contributing to extremely detrimental, systemic events. In such circumstances transfers of wealth based on technology are often evident, as less technologically advanced traders or smaller market participants prove vulnerable to the rapid and sophisticated trading strategies employed by HFT firms, inadvertently financing the profits of their more advanced counterparts.

The analysis also shows that modern trading system act as a positive economic force within financial markets, as they foster continuous innovation. Such innovation is mandatory, given the short-lived nature of dominant positions in modern trading industry, and HFT firms heavily invest to gain and maintain their technological edge and competitive capabilities. Unlike other economic sectors, the barriers to enter the industry are relatively low, along with a high probability to become suddenly outdated. Such lower entry threshold means that new competitors continuously step into the field and challenge established players, generating dynamics similar to the ones of a perfectly competitive entrepreneurial environment.

In a similar context, human capacity and know—how are key, as the ability to constantly innovate and refine trading algorithms ultimately depends on the quality of human capital. It may seem a paradox, but a robot-dominated industry is always based on advanced, rare, sophisticated, human skills.

To conclude, modern trading systems are a clear expression of a huge progress in the evolution of economic and financial markets. Their presence has changed equilibria across players and altered profit opportunities. While such systems have rapidly become a dominant "species" in the trading ecosystem, their dominance pertains to advanced trading technologies as a whole rather than to individual entities, which remain subject to intense competitive pressures, even more than other market participants. Over time, as in any natural ecosystem, the market environment has adapted to the presence of these new players, initially perceived as disruptive. Other market participants are actively gaining the necessary expertise to navigate the new landscape, trying to maximize the advantages related to the presence of modern trading systems while minimizing the risks of being adversely affected by their activity.

References

Angel, J. J., Harris, L. E., & Spatt, C. S. (2010). Equity trading in the 21st century. *The Quarterly Journal of Finance, 1*(01), 1–53.

Benos, E., & Sagade, S. (2012). *High-frequency trading behaviour and its impact on market quality: Evidence from the UK equity market* (Bank of England Working Paper No. 469).

Bessembinder, H. (2003). Trade execution costs and market quality after decimalization. *Journal of Financial and Quantitative Analysis, 6*(3), 233–257.

Biais, B., Foucault, T., & Moinas, S. (2015). Equilibrium fast trading. *Journal of Financial Economics, 116*(02), 292–313.

BIS Markets Committee. (2018). *Monitoring of fast-paced electronic markets.*

Black, F. (1976). *Studies of stock price volatility changes*. Proceedings of the American Statistical Association, Business and Economic Statistics Section.

Brogaard, J. (2010). *High frequency trading and its impact on market quality*.

Brogaard, J., Hendershott, T., & Riordan, R. (2014). High-frequency trading and price discovery. *The Review of Financial Studies*.

Chaboud, A., Rime, D., & Sushko, V. (2023). *The foreign exchange market* (BIS, Working Papers N. 1094).

Chordia, T., Roll, R., & Subrahmanyam, A. (2001). Market liquidity and trading activity. *Journal of Finance, 56*(02), 501–530.

Commodity Futures Trading Commission (CFTC) and Securities and Exchange Commission (SEC) Joint Report. (2014). *The U.S. Treasury Market on October 15, 2014: A flash crash and its aftermath*.

Degryse, H., De Winne, R., Gresse, C., & Payne, R. (2020). *Cross-venue liquidity provision: High frequency trading and ghost liquidity* (ESMA Working Paper N. 4).

Easley, D., López de Prado, M. M., & O'Hara, M. (2012). The volume clock: Insights into the high-frequency paradigm. *Journal of Portfolio Management, 39*(01), 19–29.

Flora, M., & Renò, R. (2020). *V-shapes*. University of Verona.

Foucault, T., & Moinas, S. (2018). *Is trading fast dangerous?* (Toulouse School of Economics, Working Paper).

Grossman, S. J., & Stiglitz, J. E. (1980). On the impossibility of informationally efficient markets. *American Economic Review, 70*(03), 393–408.

Hasbrouck, J., & Saar, G. (2013). Low-latency trading. *Journal of Financial Markets, 16*(04), 646–679.

Hendershott, T., Jones, C., & Menkveld, A. (2011). Does algorithmic trading improve liquidity? *Journal of Finance, 46*, 179–207.

Jovanovic, B., & Menkveld, A. J. (2016). Middlemen in limit order markets. *Annual Review of Financial Economics*.

Kirilenko, A. A., Kyle, A. S., Samadi, M., & Tuzun, T. (2017). The flash crash: The impact of high-frequency trading on an electronic market. *Journal of Finance, 72*(03), 997–98.

Lewis, M. (2015). *Flash boys. A Wall Street revolt*. W. W. Norton & Company.

Lo, A. W., Mamaysky, H., & Wang, J. (2000). Foundations of technical analysis: Computational algorithms, statistical inference, and empirical implementation. *Journal of Finance, 55*(4), 1705–1765.

Menkveld, A. J. (2013). High frequency trading and the new market makers. *Journal of Financial Markets, 16*(04), 712–740.

Menkveld, A. J., & Yueshen, B. Z. (2019). The flash crash: A cautionary tale about high-frequency trading. *Management Science*.

Panzarino, O., Potente, F., & Puorro, A. (2016). *BTP futures and cash relationships: A high frequency data analysis* (Banca d'Italia, Working Paper No. 1083).

Xu, Z., & Xue, C. (2021). Are large traders harmed by front-running HFTs? *Information Technology and Quantitative Management*.

Chapter 10
The Risk-Neutral Probability in Finance: An Epistemic Insight with a Toy Model

Christian Walter

Abstract The so-called "risk-neutral probability" is a technical tool that has received considerable attention in financial practices over the past thirty years, to the point that risk-neutral pricing techniques are now a routine part of the day-to-day business in the finance industry. I present an extremely simplified toy model of risk-neutral pricing to allow the black box to be opened to philosophical investigation by displaying without any mathematics the fair value pricing mechanism in a complete arbitraged free market. I argue that the difficulties of the puzzling methods used to value financial assets using risk-neutral valuation techniques are more conceptual than mathematical, and discuss the epistemological issues enlightened by this toy model, particularly in terms of narratives. The toy model reveals how the Efficient Markets Hypothesis (EMH) is a narrative imposed on the real finance, and how the EMH narratives correspond to each other as quantification conventions which draws a "representation format" of EMH. It raises the ethical question of choosing the EMH narrative for financial practices.

Keywords Risk-Neutral pricing · Efficient market hypothesis · Fair value · Quantification convention · Representation format

10.1 Introduction

The so-called "risk-neutral probability" is a technical tool that has received considerable attention in financial practices over the past thirty years, to the point that risk-neutral pricing techniques are now a routine part of the day-to-day business in the finance industry. The risk-neutral probability is a probability measure in a dual world, in general written down as Q in the financial literature, in distinction to the "real" (or "physical") probability of the phenomenon under consideration,

---

C. Walter (✉)
LAP (EHESS), Af2i (Association Française des Investisseurs Institutionnels), Paris, France
e-mail: christian.walter@ehess.fr

E. Ippoliti et al. (eds.), *Philosophy and Finance: Ten Open Questions*, Synthese Library 504, https://doi.org/10.1007/978-3-031-96510-4_10

denoted P. The nature of probability Q is well understood in the mathematical theory of finance. There are numerous textbooks which provide the detailed technical background to this (Cochrane 2001; Dothan 1990; Föllmer & Schied 2002; Karatzas & Shreve 1998; Pliska 1997). The shift from P to Q has been extensively described in mathematical finance with the key role of the Radon-Nikodyn operator $L = Q/P$ known in financial literature as the "state price density" (SPD). Mathematical finance has made extensive use of this technique, via the Girsanov theorem. In the case of the standard model of price fluctuations using the Brownian representation for price dynamics (Walter 1996, 2001, 2019), since Brownian motion is a continuous stochastic process, Girsanov's theorem enables the probability to be changed from the "physical" world (P) to the "dual" world built by the risk-neutral probability (Q). As shown by Harrison and Kreps (1979), the SPD exists if the market is arbitrage-free (absence of arbitrage opportunity, AOA), and it is unique if the market is complete. This uniqueness is the core condition for obtaining the "fair value" of any asset or liability and, reciprocally, a value is assumed "fair" if and only if the SPD is unique. Hence there is a "va-et-vient" between the change of probability and the notion of fair value. The risk-neutral probability Q became a convenient way of determining the "fair value" of given assets or liabilities.

In financial economics, fair value is the financial consequence of the notion of the so-called "Efficient Market Hypothesis" (EMH) alternatively known as the Efficient Market Theory. If EMH holds, it is possible to state that the market price of a given asset is "fair". With EMH, one can hypothesize that the market is a "good" mechanism for pricing any asset or liability. The history of financial thought has deeply examined the formation and the development of EMH (for a recent philosophical approach, see Walter 2017). EMH has been extensively discussed since its introduction by Eugene Fama in 1970 (Fama 1970). In particular, the joint hypothesis problem is now well identified (Jarrow & Larson 2012), which raises issues for Hempel's criteria for the logic of test of a hypothesis in the case of auxliary hypotheses (Hempel 1966, p. 22). To be falsified (Popper), EMH must be tested jointly with a given model for expected returns (Fama 1991) and another given model for expected risk (Walter 1996).

A large number of mathematical models exist to portray the situation of an informationally efficient market, as alternative representations of EMH. All these "EMH-models" have been widely documented. There is a lot of discussion and controversy about these models. In general, one works on rather complex models, and controversies do not ignore this complexity. This complexity is necessary for the accuracy of the models. It is useful in order to better apply a selected EMH-model to a given business context (e.g. among others Boyle and Feidhlim 2001; Habart-Corlosquet et al. 2013, 2015; McNeil et al. 2015; Ottaviani 1995; Reghai 2015) in a specific "culture of models" (Pickering 1992) with a pratice-oriented approach to modelling for decision-making (Svetlova & Vanessa 2014). The representation of a given mathematical framework plays a key role in testing EMH and in decision making (asset valuation, risk metrics, portfolio management).

Two main mathematical frameworks exist for mathematising EMH, I name the P-world and the Q-world. These two frameworks have been characterised in Chiapello and Walter (2016) as two "quantification conventions" (Desrosières 2003) of the real finance, in the sense that a quantification convention for financiers is a metrology that allows them to work easily on a day-to-day business. These two quantifications are respectiveley the "mean-variance convention" in the P-world and the "market consistent convention" in the Q-world. The periodisation introduced in Chiapello and Walter (2016) started with the actuarial convention, i.e. before the development of modern financial theory and EMH. Here I consider only the period starting after the birth of the financial theory and EMH. This theory evolved through two quantifications, from the P-world to the Q-world. Hence, these two quantifications conventions act as the first and second quantification of the EMH. In this paper, the P-world is named the first quantification convention of EMH and the Q-world is named the second quantification convention of EMH. The paper expands and elaborates on Chiapello and Walter (2016) by adding to it a simplified integrated approach to these two conventions, exhibiting the correspondence between these two worlds.

The fact that the probability Q is widely used in financial markets for pricing derivatives is well known. It is also well known that Q is the cornerstone of all the techniques of contemporary finance. For example, Accounting Standard Codification ASC paragraph 718-10-55-16: "Established principles of financial economic theory represent fundamental propositions that form the basis of modern corporate finance for example, the time value of money and risk-neutral valuation". It is not as well known that it is also used in the framework of financial regulation. In fact, a very interesting aspect of the practice of using probability Q is that this mathematical concept is fully incorporated into a prudential regulatory framework. The calculation technique under Q is the backbone of financial regulations implementing the MCV such as the European Solvency II Directive. Hence, the use of Q has also become a reality in areas seemingly distant from the derivatives markets. For example, a "market consistent value" (MCV) of an asset or liability is its market value, "if its is readily traded on a market at the point in time that the valuation is struck and (...) a reasoned best estimate of what market value would have been had it been readily traded at the relevant valuation point" (Kemp 2009). Given the importance of Q in the regulatory framework of finance, it seems that Q is a key part of the machinery of finance if we look at the "view from inside" of finance (Ippoloti 2017, p. 121) namely rules, laws, institutions, and regulators. The understanding of how the SPD works is important to the accurate understanding of the financial system.

Given the role that Q played in the transition of finance from the 1960s to the 1990s and in the regulatory framework, as well as in the possible financialisation of the economy, it would be important to be able to explain the stakes of the changes of worlds without needing to use sophisticated mathematics. But in this literature, there are many mathematical barriers to a simple approach to understanding the change of probability. In most cases, asset pricing models are expressed in the language of continuous-time stochastic differential equations using a martingale approach.

The richness of these models allows for a general formula but the complexity of these models is high. It could be useful to simplify these complex models by constructing a simplified artificial world. Hence, my aim here is to use a elementary static "mini-model" for risk-neutral pricing allowing to illuminate with a extremely simple approach the correspondence between the P-world and the Q-world, as well as the causes of the uniqueness of Q with AOA condition. This minimalist view allows to cognitively grasp the relations between the two frameworks P and Q and their mathematical representations, the P-world and the Q-world, without any mathematical complexity. The paper is organised as follows.

Section 10.2 introduces the mini-model and the economy. Section 10.3 uses the mini-model and describes the arbitrage mechanism. I present the arbitrage effect on prices, the change of probability and why the probability Q becomes unique after arbitrage. Section 10.4 presents the different representation formats of the EMH with their discounting convention, their mathematical correspondence and the link between CAPM, risk-neutral probability Q, market pricing kernel and Arrow-Debreu securities in the mini-model. Section 10.5 concludes and opens the debate by presenting the epistemic contributions of the mini-model for non-specialists and financial practitioners.

10.2 The Toy Model

The static toy model introduced here is a mini "EMH-model". It allows to deeply understand the idea of EMH in a extremely simple artificial case.

10.2.1 Extremely Simple Models

From an epistemological standpoint, an highly idealized and extremely simple model is sometimes called a "toy model" (Reutlinger et al. 2018). According to these authors, there are two kinds of toy models, named "embedded toy model" and "autonomous toy model" (Reutlinger et al. 2018). An autonomous toy model is designed without any background theory, as for instance the canonical Schelling's checkerboard model for segregation (Schelling 1971). An embedded toy model is a that of "empirically well-confirmed framework theory". According to Reutlinger et al. (2018), the mini-model I present here is an embedded toy model. The theories I use are that of the framework of mean-variance optimisation introduced by Harry Markowitz (1952) for portfolio management, leading to the Capital Asset Pricing Model (CAPM) of Sharpe (1964), and the framework of risk-neutral pricing introduced by Harrison and Kreps (1979) and Harrison and Pliska (1981).

From a mathematical standpoint, I use a well-known framework in financial modelling since the Cox et al. (1979) model, the binomial one, including two risky assets, one money market and two scenarios for the future. The relevance of the binomal framework is very known today. This framework is proving very useful in explaining the functioning of the pricing methods. With respect to the existing literature on the binomial framework in finance and the very numerous examples that can be found on the field, my aim here is to carry this framework as far as possible by using it to the point of highlighting the values of the Radon-Nikodyn operator L in a static model with a binomial framework. The static mini-model allows to grasp the signification of the Radon-Nikodyn function without mathematical expertise, enlightening the interplay between these two mathematical representations that are used in day-to-day financial practices. The novelty in the paper lies in the way I demonstrate the connection between the representations of the two worlds ("real" and "dual") achieved by presenting numerical values and linking them with the CAPM numbers

From an educational standpoint, the proposed mini-model is a piece which provides access to a theoretical overview of the mathematical scene of neoclassical finance to non-experts in mathematical finance and stochastic calculus. It provides an easy epistemic access to understanding the key role of the risk-neutral probability Q in work related to asset pricing models. Hence my ambitiously stated proposal is that the epistemic goal of my mini-model is to enable non-specialists of stochastic calculus ("non-quant" finance practitioners or researchers in the humanities and social sciences) to understand some of the epistemic difficulties of the risk-neutral pricing, namely the intertwin between the P-world and the Q-world and the uniqueness of Q with AOA in complete markets.

A possible objection to the relevance or usefulness of this mini-model is that an extremely simplified model is too simple or too abstract to be explanatory (Cartwright 2009). But many scientists consider that an abstract and simplified model sheds light on important aspects that are often obscured by the current use of complex models, even if the simplifications make the model highly unrealistic (Ylikoski & Aydinonat 2014). Abstract and simplified idealized models have an important role in many sciences and creating an stylized world can contribute to the scientific understanding of complex concepts. Here, the mini-model allows to simply but deeply understand the strange mechanism of uniqueness of Q after arbitrage in complete market, it allows to debunk the mathematical trick which is at the basis of any pricing in modern finance.

I now present the mini-model and its use to exhibit extremely simply the uniqueness of risk-neutral probabiliy Q with AOA. The uniqueness of Q emerges from a rule-of-three calculation, without mathematical puzzles. The two values of the Radon-Nikodyn operator L (state price density) are calculated in this very simple framework. The epistemic value of the quotes has changed after arbirage.

10.2.2 The Economy

Let us begin with the information set. Fama (1970)'s celebrated review about EMH divided the notion of market efficiency into three categories: (1) weak-form (tests to scrutinize if past returns predict future returns), (2) semi-strong-form (tests to check if security prices reflect public information announcements), and (3) strong-form (investigation if investors have private information that is not fully reflected in market prices). For the simplicity of my purpose, I now assume that all the available information is used to elaborate scenarios for future values of assets. Note that, instead of information about companies, it could be possible to use data sets as information, following the case-base decision theory (Gilboa & David 2012).

It should be noted that there is no guarantee that the forecasted future values have any connection with the economic situation of the companies. It would be equally possible to imagine future forecast values that are not based on financial analysis calculations, but on rumours about potential developments in the companies, without any business scenario calculations. To put it differently, the forecasts can be either "rational expectations" or any kind of rumours or "noise". The agents can be rational investors, noise traders or chartists. For the purpose of the mini-model, at this stage, this distinction is not crucial.

Let us continue with the description of uncertainty. We assume that uncertainty of the outcome is measurable and, following Frank Knight (1921, for recent debates see Diebold et al. 2010, Svetlova 2021 and Dimand 2021), we use the term "risk" to designate measurable uncertainty. That means that the probability distribution of the outcome is known, "either through calculation *a priori* or from statistics of past experience" (Knight 1921, p. 233). We use the Arrow and Debreu (1954) framework of a general equilibrium under uncertainty. Uncertainty about the outcomes of an action is described as choice over state-contingent outcomes by a finite set of well-defined and perfectly known states of the world representing the known "unknowns" denoted Ω with ω being a generic element. An observable subset of Ω is called an "event". It can be a single state or a collection of states. The scenarios for the contingent futures can be thought as possible states or "events" of the future. It is well known that this theory was extended by the subjective probability distribution introduced by Savage (1954) and criticised by Ellsberg (1961) and later Kahneman and Tversky (1979) to take account of the many puzzles raised by the Arrow-Debreu approach. Nevertheless, in order to address issues relating to EMH and the mini-model without too much complexity, we will remain here within the classic framework. We can, nonetheless, note that this extended framework makes it possible to interpret the risk-neutral probability Q as the objectification of an implicit and shared subjective measure of the market.

Let us imagine a company A for which financial analysts have elaborated at date $t = 0$ scenarios for a given time horizon, say $t = 1$. It is possible to understand these scenarios as subjective beliefs of analysts, i.e. subjective predictions about the occurrence of the states the world. For the simplicity of the model, we suppose there are only two scenarios, growth or decline, correponding to pessimistic forecasts and optimistic forecasts. The set Ω of scenarios is ("down", "up") written as $\Omega = \{\omega_d, \omega_u\}$. On the date $t = 0$ of the valuation, the share price of company A quoted on the stock exchange (or its estimated value if the company is not quoted on a stock exchange) is $A_0 = 150$ euros. In the first scenario (an optimistic version of the future of the company), the value of the share at $t = 1$ would be, for example,

$A_1(\omega_u) = 200$ euros. In the second scenario (more pessimistic about the business prospects), the share value would fall to $A_1(\omega_d) = 140$ euros.

Let $P(\omega)$ be the probability of the scenario ω. The probability P is the probability of the "phenomenon" related to the future of the company, the future state of "event" of the "physical world" or "real world". The chosen framework guarantees that the occurrence of future events can be calculated with probabilities. In this conceptual framework, the future share value $A_1(\omega)$ is a random variable depending of the event ω. Suppose that neither scenario is preferred to the other: the two business prospects are assumed to be equally likely, and the probability of each scenario is therefore $P(\omega_u) = P(\omega_d) = 0.50$. Of course, this probability could be different (see Knight above) and any other values would be possible as long as the sum of the probabilities of each scenario remains equal to 1. The two probabilities of 0.50 applied to the two forecast values of 140 and 200 euros represent the probability distribution of the future value of the company. Since there are two possible outcomes for the future value of the shares, each weighted by a probability, it is termed a "risky prospect". The word "risky" means that the uncertain future is calculated with probabilities.

Let us now enlarge our example by considering the "economy". To keep the toy nature of the model and to preserve the mathematical simplicity of the example, we limit the economy to two companies. There is a second company denoted B which, like the first, is the subject of financial forecasts. Let us imagine that the share price at $t = 0$ is $B_0 = 66$ euros. In the first business scenario, at $T = 1$ the share value would be, for example, $B_1(\omega_d) = 50$ euros. In the second scenario, it would rise to $B_1(\omega_u) = 120$ euros. The market is complete. The probability of the scenarios is the same as for company A. It exists a money market in the economy. This market provides a risk-free rate, i.e. a rate of return whose value does not depend on the scenarios, unlike the risky prospects. That means that, whatever the scenarios, the future value of the money market asset will be the same. The risk-free rate is equal to 2%.

Table 10.1 summarises the "mini-economy" with both risky prospects. Table 10.2 present the money market. In Table 10.1, the matrix [140, 200 ; 50, 120] represents

Table 10.1 The price system of the mini-model of economy before arbitrage

Real world probability P		**0.50**	**0.50**
Scenario		Pessimistic	Optimistic
Type of the asset	Asset price at $t = 0$	Forecast for $T = 1$ (epistemic values)	
A-company share	150	140	200
B-company share	66	50	120

Table 10.2 The money market of the mini-model of economy

Real world probability P			**1**
Type of the asset	Asset price at $t = 0$	Risk free rate	value for $T = 1$
Money market risk free rate	1	2%	1.02

the forecasts based on the information or rumours avaiblable to the participants of market, yielding the price system [150 ; 66] . In this sense, [140, 200 ; 50, 120] are the "epistemic values" of the quotes [150 ; 66]. In Table 10.2, a risk-free asset of value 1 at $t = 0$ produces 1.02 at $T = 1$.

10.3 The Toy Model at Work

We now use the mini-model to show in an extremely simple way how arbitrage in a complete market generates the uniqueness of the risk-neutral probability Q.

10.3.1 Trading Strategies and Financial Arbitrage

In the mini-model, a trading strategy is an asset allocation of funds between the two companies A and B. It is possible to define this portfolio with number of shares or weight of companies.

The notion of arbitrage is crucial in finance and financial theory. It is the cornerstone of the option pricing theory due to Black and Scholes (1973) but more generally for pricing any asset in practice. In finance, a financial arbitrage is a trading strategy that takes advantage of price inefficiencies, i.e. a trading strategy that allows obtaining a sure gain without bearing the slightest risk.

10.3.1.1 Searching for Arbitrage

Is there an arbitrage opportunity in this mini-economy? In general, a market is screened for arbitrage by exploring the possibilities of obtaining a sure (non-random) return using a mix of risky prospects bearing random returns. In the mini-model presented, a trading strategy will consist on an asset allocation between the two companies A and B that allows obtaining a sure return. If there are prices inefficiencies in the price system (150, 66), there will be opportunities to obtain a sure return greater than the risk-free rate (2%) with an appropriate portfolio.

Let us imagine that we now seek such a trading strategy. We aim to build a portfolio of A and B such that, whatever the future scenarios, the value of this portfolio is 100,000 euros at date 1 whatever the future state of the world, i.e. it will be "risk-free". We have to find the holdings of companies A and B. Let θ_A and θ_B be the quantities ("holdings") of the shares A and B at date 0. Given the price system (150, 66), the value of this portfolio at date 0 is:

$$V_0 = \theta_A \times 150 + \theta_B \times 66 \tag{10.1}$$

At date 1, the value of this portfolio is depending on the scenario ω:

$$\theta_A \times A_1(\omega) + \theta_B \times B_1(\omega) = V_1(\omega) \tag{10.2}$$

We want V_1=100,000 whatever the scenario ω is. This leads to a system of two equations:

$$\begin{cases} \theta_A \times 140 + \theta_B \times 50 &= 100,000 \\ \theta_A \times 200 + \theta_B \times 120 &= 100,000 \end{cases} \tag{10.3}$$

whose solution is $\theta_A = 1,029.41$ and $\theta_B = -882.35$. The negative sign of θ_B means that we have to "short" the shares of company B. A short position is created when a trader sells a security first with the intention of repurchasing it at the end of the arbitrage. One can easily check that whatever the possible outcomes, the portfolio will be valued at 100,000 euros on date 1:

$$\begin{cases} 1,029.41 \times 140 - 882.35 \times 50 &= 100,000 \\ 1,029.41 \times 200 - 882.35 \times 120 &= 100,000 \end{cases}$$

At date 0, following (10.1) the value of this portfolio is:

$$V_0 = 1,029.41 \times 150 - 882.35 \times 66 = 96,176.47$$

The return of this portfolio is: $100,000/96,176.47 - 1 = 3.98\%$ without any uncertainty about the result. For this reason, the return of 3.98% is said "risk-free". Such a situation is very interesting: it would be enough to borrow from the money market (2%) the amount needed to buy this portfolio, to obtain the sure return (3.98%) without risk, and then to cash in the difference. It is therefore an anomaly in the market, since it would be possible to make very large gains without taking any risk: just buy very large quantities of A, borrow the money to finance the purchase, and short as much B as necessary. With $\theta_A = 1,029.41$ and $\theta_B = -882.45$, the resulting gain is 1,900 euros. That means that you can make 1,900 euros with zero euros spent, a kind of "free lunch".

Such an anomaly represents an inconsistency in the price formation according to the target prices forecasted in the case of EMH. Following Fama (1970), that means it exists some information which is not reflected in the price system (150,66) at date $t = 0$. This inefficient situation cannot last long because sooner or later someone (a trader, an investor, an arbitrageur) will detect it and intervene in the market to take advantage of it. By doing so, he will eliminate the inefficiency so that, at the end of the trading strategy, the market will be "cleared" of all possible arbitrage. To put it differently using the financial jargon, the condition of "absence of opportunity of arbitrage" (AOA) will be satisfied. Let us now move inside the mechanism of arbitrage, the principle that produces the market equilibrium, that is to say, what steps are required to produce the equilibrium.

10.3.1.2 Implementing Arbitrage and Effect on Prices

The effect of the arbitrage is to move prices. Buying 1,029.41 shares of company A on the market will have the effect of putting upward pressure on the price, while selling 882.35 shares of company B will have the opposite effect. For example, let us imagine that the new price system is (151.25, 65.75). The implied returns resulting from θ_A and θ_B with this new price system would be $2.37\% > 2\%$. The market is not arbitraged since it is possible to find a positive risk-free rate that allows for a sure gain without an initial capital payment.

What would be the right equilibrium price level? By toying with the model, we realize that we have to do a specific hypothesis on the price changes. Suppose that only the price of company B changes. This may mean that the information on company B was not fully taken into account in the first valuation. In this case, company B was yielding too low a return relative to company A. The arbitrage will have the effect of increasing the return of company B. The price of company B will fall until its new value is low enough that any arbitrage no longer yields a definite return higher than that of existing risk-free investments.

The arbitrage mechanism is the following. Keeping in mind that:

$$1029.41\ A_1 - 882.35\ B_1 = 100{,}000$$

whatever the state of the economy at date $t = 1$, we should therefore expect to have at $t = 0$ a return of 2% corresponding to 98,039.22 euros:

$$1029.41\ A_0 - 882.35\ B_0 = \frac{100{,}000}{1.02} = 98,039.22$$

Only the price of company B changes. In this case, the price adjustment mechanism would lead to:

$$B_0 = \frac{1}{882.35}\left(1029\ A_0 - 98{,}039.22\right) = 63.89$$

We find that the value of B after arbitrage is $B_0 = 63.89$ euros. Through the intervention of arbitrageurs, prices have readjusted to a new price level, leading to the new price system of Table 10.3. The valuation of B at the new price 63.89 is called "market consistent valuation" in the regulatory framework (Wütrich et al. 2015).

We notice that arbitrage is always relative and does not indicate what the return of company B "should be" from an economic point of view. It is just a test of the consistency of investors' expectations on the two companies simultaneously, whether these expectations are economically sound or not. If company A was different, another coherence would have arisen from the market with the couple (A,B).

Table 10.3 The price system of the mini-economy before and after abitrage

Real world probability P			**0.50**	**0.50**
Scenario			Pessimistic	Optimistic
Type of the asset	Asset price at $t = 0$		Forecast for $T = 1$ (epistemic values)	
A-company share	150.00	150.00	140	200
B-company share	66.00	**63.89**	50	120
Arbitrage	Before	After		

10.3.2 The Change of Probability

Any asset pricing model stems from one simple concept: price equals expected discounted payoff. This leads to the present value equation.

10.3.2.1 The Present Value Equation

The expected payoff of company A at date $t = 1$ is $\mathbf{140} \times 0.50 + \mathbf{200} \times 0.50$. It follows that the share value 150 at date $t = 0$ is generated by the discounting of expected payoff at date $t = 1$. One can write the "present value equation" for company A:

$$150 = \frac{\mathbf{140} \times 0.50 + \mathbf{200} \times 0.50}{1 + 0.1333} \tag{10.4}$$

The rate of 13.33% is the discount rate used to equate the expected future values at date $t = 1$ and the quoted price at date $t = 0$.

In the case of company B, the same calculation would yield the following present value equation:

$$66 = \frac{\mathbf{50} \times 0.50 + \mathbf{120} \times 0.50}{1 + 0.2879} \tag{10.5}$$

The rate of 28.79% is the discount rate used to equate the expected future values and the quoted price.

10.3.2.2 The Present Value Under *P* and Under *Q*

Now imagine that the beliefs about the probabilities of future scenarios are not the same. Instead of having an equally weighted probability distribution (50, 50), imagine that the beliefs overweight the pessimistic scenario, for example (70, 30). Considering the present value equation, there are two options: change the share price at date 0 or change the discount rate. If the share price is fixed, the present value equation becomes:

$$150 = \frac{\mathbf{140} \times 0.70 + \mathbf{200} \times 0.30}{1 + \text{new discount rate}} \tag{10.6}$$

It is straightforward to see that these constraints give the value of 5.33% for the new discount rate. We see that, given the initial share value and the expected scenarios, there is a relationship between the probability distribution and the expected rate of return.

But it can also be considered that, in this case, the expected rate of return of 5.33% for company A is not the "right" one, in the sense that it does not reflect what can be expected of a return in this economic area. The next question could be: if a rate of return is specified, what should be the probabilities of the scenarios? The answer is given with the following equation:

$$150 = \frac{\mathbf{140} \times p + \mathbf{200} \times (1 - p)}{1 + \text{specified discount rate}} \tag{10.7}$$

We can see that, once the discount rate is fixed, the probability values follow.

Let us now imagine that we decide on the expected return of company A as the risk-free rate of return, 2%. This unusual situation (because company A is risky for the investor, it is a risky prospect) would mean that the investor does not demand a risk premium for buying company A. This would mean that the investor is not affected by risk-taking, that he is "risk neutral". In this case, the present value equation becomes:

$$150 = \frac{\mathbf{140} \times p + \mathbf{200} \times (1 - p)}{1 + 0.02} \tag{10.8}$$

It is straightforward to see that this constraint gives the value of 0.7833 for the new probability, so that:

$$150 = \frac{\mathbf{140} \times 0.7833 + \mathbf{200} \times 0.2167}{1 + 0.02} \tag{10.9}$$

This new probability distribution (0.7833, 0.2167) is called the "risk-neutral probability" for company A. This new probability depends on company A, as its values are extracted from the market calibrated to the quoted price (or value) of company A. For this reason, it is denoted Q_A with the subscript A. The price 150 is now a present value in a "Q_A-world". The informational market equilibrium is written here in the "dual" world using the probability Q_A in the asset pricing equation.

But, it could be argued, risk aversion has not vanished from the psychology of investors. We can see that the risk aversion has been mathematically transferred from the risk premium to the new probability distribution, from the denominator of the present value to the numerator. In the Q_A-world, the risk aversion is present in the new probability distribution.

Table 10.4 The Q values in the mini-economy before arbitrage

Real world probability P		**0.50**	**0.50**
Scenario		Pessimistic	Optimistic
Type of the asset	Asset price at $t = 0$	Forecast for $T = 1$ (epistemic values)	
A-company share	150	140	200
B-company share	66	50	120
Risk neutral probability Q_A		**0.7833**	**0.2167**
Risk neutral probability Q_B		**0.7526**	**0.2474**

In the case of company B, the same calculation would yield the following present value equation:

$$66 = \frac{\mathbf{50} \times 0.7526 + \mathbf{120} \times 0.2474}{1 + 0.02} \tag{10.10}$$

The probability $Q_B = (0.7526, 0.2474)$ resulting from risk neutral constraint on company B is not the same as the probability $Q_A = (0.7833, 0.2167)$ resulting from risk neutral constraint on company A. Thus, the constraint of calculation with risk neutrality leads to as many Q distributions as there are companies. In the mini-model with two companies, we find two risk neutral probabilities Q_A and Q_B (Table 10.4).

10.3.2.3 The Uniqueness of Q After Arbitrage

Again, one might argue, what is the point of calculating a present value that gives the same result but with a different probability if, in addition, this probability depends, like risk premiums and betas, on each company? The answer is given by the transformation of prices after arbitrage. Once the market is arbitraged, the probability Q becomes unique.

With the arbitraged price of company B, the present value calculation yields the new present value equation:

$$63.89 = \frac{\mathbf{50} \times 0.50 + \mathbf{120} \times 0.50}{1 + 0.3304} \tag{10.11}$$

The rate of 33.04% is the discount rate used to equate the expected future values and the quoted price in the arbitraged market.

If we now recalculate the probability Q_B from the equilibrium price of B after arbitrage, we find that Q_B (pessimistic scenario) = 0.7833 and Q_B (optimistic scenario) = 0.2167, hence we get the following:

$$63.89 = \frac{\mathbf{50} \times 0.7833 + \mathbf{120} \times 0.2167}{1 + 0.02} \tag{10.12}$$

Table 10.5 The market equilibrium after arbitrage: the AOA condition

Real world probability P		**0.50**	**0.50**
Scenario		pessimistic	optimistic
Type of the asset	Asset price at $t = 0$	Forecast for $T = 1$ (epistemic values)	
A-company share	150	140	200
B-company share	63.89	50	120
Risk neutral probability $Q_A = Q_B = Q$		**0.7833**	**0.2167**
State price density $L = Q/P$		**1.5666**	**0.4334**

Let us recall that Q_A (pessimistic scenario) = 0.7833 and Q_A (optimistic scenario) = 0.2167. We observe that, after arbitrage, $Q_A(\omega) = Q_B(\omega)$ for each scenario ω. Let us denote Q the unique risk-neutral probability of the arbitrage-free economy. The uniqueness of the probability Q derived from the arbitraged market is due to the market is complete. This uniqueness allows both companies A and B to be valued with the same discount rate, the risk-free rate. This is the "risk neutral valuation" or "risk neutral pricing". The uniqueness of the probability Q is closely linked to the absence of arbitrage, the "no arbitrage condition" ("absence of arbitrage opportunity", AOA). The risk-neutral probability Q is the mathematical "natural child" of the EMH.

The change in probability measure is the cornerstone of rational pricing. This calculation technique is used to transform "real world" probability P into risk-neutral probability Q. Once the uniqueness of Q is obtained after arbitrage if the market is complete, all variables relevant to financial calculations (pricing, hedging, position fixing etc.) are requantified. How this requantification works is the role of the state price density $L = Q/P$, the mathematical operator which transforms the P-world into the Q-world. We can see that moving from P to Q is equivalent to overweighting the pessimistic scenario by $0.7833/0.5000 = 1.5666$. More generally, we can calculate the ratios of the shift from the P-world to the Q-world for the probabilities of each scenario: Q (pessimistic)/P (pessimistic) $= 1.5666$ and Q (optimistic)/P (optimistic) $= 0.4334$. As $L(\omega) = Q(\omega)/P(\omega)$, we have L (pessimistic) $= 1.5666$ and L (optimistic) $= 0.4334$. The mini-model allows to easily calculate the values of the state price density.

Hence, the mini-model confronts the two quantifications of EMH provided by the use of P and Q in the same very tangible and simple situation, with a stylized elementary market moving towards equilibrium (Table 10.5).

10.4 The Alternative Representations of EMH

Now that we have described the mini-model and the impact of arbitrage on prices, we move on to the pricing of assets. A common definition of asset pricing is that of Cochrane (2001, p. xiii): "asset pricing theory tries to understand the prices of values of claims to uncertain payments" adding that "a low price implies a high rate of return, so one can also think of the theory as explaining why some assets pay high average returns than others".

The alternative representations of EMH are alternative asset pricing methods using a given discounting convention. We now elaborate on this.

10.4.1 The Three Discounting Conventions

At the most general level, a discount factor is just a variable that generates prices from payoffs. It has been shown in Chiapello and Walter (2016) that three "discounting conventions" have organized the history of financial pratices. The discount factor has been transformed by the emergence of each discounting convention. The mini-model allows to grasp the nature of these changes.

10.4.1.1 The Actuarial Convention

The first discounting convention is the "actuarial convention". With this convention, the present value is determined through a simple calculation: known cash flows were discounted to present value using a constant interest rate. The discount factor is deterministic.

As seen above in the present value Eq. (10.4), the share value 150 is generated by the beliefs and the discounting of possible future values:

$$150 = \frac{\mathbf{140} \times 0.50 + \mathbf{200} \times 0.50}{1 + 0.1333}$$

We now rewrite this equation by highlighting the discount factor:

$$150 = \mathbf{140} \times \frac{1}{1.1333} \times 0.50 + \mathbf{200} \times \frac{1}{1.1333} \times 0.50$$

The quantity $1/1.1333 = 0.8824$ is called the discount factor of company A. We rewrite the previous relationship by making the discount factor visible:

$$150 = \mathbf{140} \times 0.8824 \times 0.50 + \mathbf{200} \times 0.8824 \times 0.50$$

The equilibrium price 150 is the mathematical conditional expectation of the payoff given the information set, "discounted" using the discount factor 0.8824. In Fama's view (1970), this relation says that 150 "fully reflects" the available information on A.

Doing the same thing with company B, we find:

$$63.89 = \mathbf{50} \times 0.7517 \times 0.50 + \mathbf{120} \times 0.7517 \times 0.50$$

The quantity $1/1.3304 = 0.7517$ is the discount factor of company B.

Now it is possible to write the price system with company-specific discount factors:

$$\begin{cases} 150.00 = \mathbf{140} \times 0.8824 \times 0.5000 + \mathbf{200} \times 0.8824 \times 0.5000 \\ 63.89 = \mathbf{50} \times 0.7517 \times 0.5000 + \mathbf{120} \times 0.7517 \times 0.5000 \end{cases} \tag{10.13}$$

This price system (10.13) reflects the EMH with company-specific risk-adjusted discount factors: 0.8824 for company A and 0.7517 for company B.

10.4.1.2 The Mean-Variance Convention

The price system (10.13) describes the EMH with company-specific risk-adjusted discount factors. The issue is the choice or the calculation of the company-specific discount factors. In most of cases, it is possible to choose a given discount factor resulting from financial analysis of the company. While some idea of risk is empirically taken into account by the choice of a higher or lower discount rate, this risk is not based on a statistical calculation in the first discounting convention.

The main interest of the second discounting convention is to allow to find a discount factor including a risk level with the CAPM, a risk premium. In the "mean-variance convention", following Markowitz (1952), the risk is defined by the variance (or its square root, the standard deviation). The level of the risk premium is determined using the CAPM, which gives the level of the risk premium as a linear relation to the beta coefficient. Under the second discounting convention, the discount factor is no longer deterministic and becomes variable, since it depends on the beta coefficient. As discount factors are given under CAPM, we will speak of EMH-CAPM, a label which exemplifies the joint hypothesis problem.

Let us emphasise this point. The second discounting convention introduces a new and extremely important idea for the financial practices: the relevant discount rate for calculating a present value is related to the rate of return on a specific portfolio known as the "mean-variance-optimal tangent" portfolio. This portfolio has been considered equivalent to the "market" since Sharpe's (1964) seminal paper. Apart from the technicity of this concept, the new development is that valuation is now associated with market equilibrium, paving the way for the "financialisation of valuation" (Chiapello 2015). In the second discounting convention, valuation of any item requires a mean-variance optimal tangent portfolio, which in practice means actors must keep up with an index. And conversely, any mean-variance optimal tangent portfolio (or market index) becomes a possible instrument for asset valuation. Hence the "market" needs a proxy representation in order to apply the theoretical financial research to make practical real-life decisions. Serving as proxies is precisely the function of market indexes such as the Dow Jones Euro Stoxx (DJEX).

However it is well known today that a lot of non-normality features (see Cont 2001, for example) invalidate the pure mean-variance (MV) framework. To take account for this, researchers consider augmenting the classic MV optimisation with higher order moments, namely skewness S and kurtosis K parameters for porfolio selection, where skewness and kurtosis define the asymmetry and peakedness of risk profile (see for example Jondeau and Rockinger 2006, Jurczenko and Maillet 2012). More precisely, kurtosis is the parameter which measures the tail thickness of return distribution. This approach gives birth to a multi-moments approach to risk, a generalised mean-variance-skewness-kurtosis (MVSK) framework. Including S and K in the higher dimensional portfolio selection problem is a convenient way to capture extreme risk, and it would be possible to sketch an EMH-MVSK model. But with extreme risk, the exhibition of a simple Radon-Nikodyn measure L is not easy (see Le Courtois et al. 2023, for discussion and details). To not overcomplicate the mini-model, we decide to not take account of the issue of asset allocation with extreme risk and to stay in the MV framework of the mean-variance convention (see Le Courtois et al. 2016, for an example of treatment of the asset allocation problem with extreme risk).

Coming back to the mini-model in the mean-variance convention, to exhibit the EMH-CAPM representation, we need to build a "mini CAPM". To be able to do this, we have to find an equivalent of the "market portfolio", i.e. an index representative of the two companies A and B, for example, with reference to the DJ Euro Stoxx 50 (DJEX50), the "DJ Euro Stoxx 2" (DJEX2). The DJEX2 index consists of a portfolio of only companies A and B, whose weights in the portfolio are calculated using Markowitz mean-variance optimisation with the values for each scenario. Let us denote w_A^M and w_B^M the weights of company A and company B in the DJEX2. The weights w_A^M and w_B^M are the mean-variance-optimal weights of the companies in the DJEX2 index. Following Sharpe's CAPM, they represent their "market capitalisation" in the mini-economy in the mini EMH-model. Using the mini-model, we find $w_A^M = 0.7326$ and $w_B^M = 0.2674$ (details of calculation available on request).

By applying these weights to the expected returns of A and B, it is easy to deduce the Markowitz' risk-return representation of the DJEX2 index:

$$\begin{cases} \text{expected return} = 18.60\% \\ \text{expected risk} \quad\; = 16.05\% \end{cases} \tag{10.14}$$

The Table 10.6 presents the mean-variance convention and the mini CAPM resulting from the model, the risk level being characterised only by the standard deviation without any higher moment (S or K). Let us assume that the risk-free rate is 2%. The risk premium of market index is $18.60\% - 2\% = 16.60\%$.

Now we have to find the betas of A and B in the mini-model, obtained from the prices of 150 and 63.89 euros. A calculation with the mini-model yields for company A the value of 0.68 and for B the value of 1.87, respectively. This means

Table 10.6 The mean-variance convention and the mini CAPM

	Porfolio weights		Market price of risk without higher moments			
					RISK	
	A	B	Expected return **E**[.]	Risk premium $\pi(.)$	Expected volatility σ	Expected beta β
Money market			+ 2.00%	0	0.00%	0.00
Company A	1	0	+ 13.33%	+ 11.33%	20.00%	0.68
Company B	0	1	+ 33.04%	+ 31.04%	54.78%	1.87
DJEX2	0.7326	0.2574	**+18.60%**	**+16.60%**	**16.05%**	**1.00**

that if the return on the DJEX 2 index increases by 1%, the return on company A will increase by 0.68% and that on company B will increase by 1.87%. Company B amplifies market movements, while company A dampens them. These fluctuations are obviously related to the "MV-optimal market portfolio" of the CAPM. With another market portfolio or a "MVSK-optimal market portfolio", the values of the betas would be different.

We can therefore write the EMH-CAPM in the mini-model with the beta equation as follows:

$$\begin{cases} 13.33\% = 2\% + 0.68 \times (18.60\% - 2\%) \\ 33.04\% = 2\% + 1.87 \times (18.60\% - 2\%) \end{cases} \tag{10.15}$$

This system is the exact Sharpe's (1964) CAPM. We have now written the CAPM in the mini-model, i.e. we have built a "mini CAPM". There are only two equations since there are only two companies.

10.4.1.3 The Market Consistent Convention

As explained in Chiapello and Walter (2016), when a new convention arises, it does not mean the disappearance of the old one. In the case of risk-neutral probability, the actuarial discoungting convention is still used with the new probability Q. The share value 150 is generated by the beliefs and the discounting of possible future values with the new probability and the risk-free rate, so that the present value Eq. (10.4) becomes:

$$150 = \frac{\mathbf{140} \times 0.7833 + \mathbf{200} \times 0.2167}{1 + 0.02}$$

We now rewrite this equation by highlighting the discount factor:

$$150 = \mathbf{140} \times \frac{1}{1.02} \times 0.7833 + \mathbf{200} \times \frac{1}{1.02} \times 0.2167$$

The quantity $1/1.02 = 0.9804$ is called the risk-free rate discount factor. We rewrite the previous relationship by making the discount factor visible:

$$150.00 = \mathbf{140} \times 0.9804 \times 0.7833 + \mathbf{200} \times 0.9804 \times 0.2167 \tag{10.16}$$

And the same calculation for company B.

Now it is possible to write the price system with risk-free rate discount factor, a single discount factor for the two companies:

$$\begin{cases} 150.00 = \mathbf{140} \times 0.9804 \times 0.7833 + \mathbf{200} \times 0.9804 \times 0.2167 \\ 63.89 = \mathbf{50} \times 0.9804 \times 0.7833 + \mathbf{120} \times 0.9804 \times 0.2167 \end{cases} \tag{10.17}$$

The price system (10.17) reflects the EMH with risk-neutral probability Q and risk-free rate for discounting, written EMH-Q.

In the previous cases, the discount factor was constant or variable but not random. A natural generalisation of this framework is to view the discount factor as stochastic, i.e. contingent on scenarios ω_1 and ω_2 or "states of the world". The discount factor, which in the second discounting convention only varied with the investments studied (i.e. the risk specific to each one, measured by the beta), has now become random. "Stochastic discounting" replaces traditional discounting, whether the discount rate used is given (under the first discounting convention) or results from an equilibrium model such as the CAPM (under the second discounting convention). This constitutes the third discounting convention named in Chiapello and Walter (2016) the "market-consistent convention".

The notion of the "stochastic discount factor" (SDF) was introduced by Duffie (1996 [2016]). Within the general framework of stochastic discounting, the SDF method provides a unified general approach to the econometric analysis of asset pricing models. The SDF expands the notions of variable but not random discount factor and it is used to represent the pricing operator in dynamic stochastic economies. The SDF is also termed the "deflator", just as a traditional operation deflates nominal values to real values. The SDF is the another terminology of the "Market Pricing Kernel" (MPK) of Lucas (1978). In financial litterature, the terms MPK and SDF are considered interchangeable.

We now display the SDF in the mini-model. Let us denote $\alpha(\omega)$ the SDF with the ω state. The relation between the SDF and the state price density (SPD) is:

$$\alpha(\omega) = \frac{L(\omega)}{1+r} = \frac{1}{1+r}\frac{Q(\omega)}{P(\omega)} \tag{10.18}$$

This means that, for a given state of the economy, the SDF is equal to the SPD adjusted by the risk-free rate. It follows from (10.18) that:

$$\begin{cases} \alpha\,(\text{pessimistic}) = 1.5359 \\ \alpha\,(\text{optimistic}) \;\; = 0.4248 \end{cases} \tag{10.19}$$

Using the market pricing kernel, the price system at equilibrium is;

$$\begin{cases} 150.00 = \mathbf{140} \times 1.5359 \times 0.5000 + \mathbf{200} \times 0.4248 \times 0.5000 \\ 63.89 = \mathbf{50} \times 1.5359 \times 0.5000 + \mathbf{120} \times 0.4248 \times 0.5000 \end{cases} \tag{10.20}$$

The price system (10.20) reflects the EMH with MPK, written EMH-MPK.

10.4.2 Dialogue with the General Equilibrium Theory

We now engage in a dialogue with general equilibrium theory of Arrow (1959) and Gérard Debreu (1959) by highlighting the relationship between the EMH and this theory. The idea of a complete market means that all financial products can be replicated by a linear combination of elementary assets, as "elementary particles" of all financial products, the so-called "Arrow-Debreu securities". The value of the discount factor for each different state of the market is the price of the Arrow-Debreu security quantifying this particular state adjusted by the probability of the state. For this reason, the completeness of the market and the uniqueness of the risk-neutral probability Q are like two sides of the same coin, the game of fortune in an arbitraged market, shadowed by the Arrow-Debreu securities.

For this purpose, we consider the market states (the two scenarios) as "states of the world" in the sense of the general equilibrium model of Arrow and Debreu. We now introduce the notion of Arrow-Debreu security. An Arrow-Debreu security, also known as a pure security or a primitive asset, is a contract that pays one unit of cash (a currency or commodity) if a particular state of affairs occurs at a particular time in the future and pays zero cash in all other states. These fictional assets are the fundamental elements on the basis of which economic equilibrium in an uncertain environment is currently described.

In the static mini-model, let us note e_t^k for $t = 0, 1$ the price at date t of the elementary asset giving to its holder an income of 1 at date 1 if the event k =pessimistic or optimistic occurs. It follows from the definition of an Arrow-Debreu security that the prices of the Arrow-Debreu securities in the mini-model are:

$$\begin{cases} e_0^{\text{pessimistic}} = \frac{Q(\text{pessimistic})}{1.02} = \frac{0.7833}{1.02} = 0.7680 \\ e_0^{\text{optimistic}} = \frac{Q(\text{optimistic})}{1.02} = \frac{0.2167}{1.02} = 0.2124 \end{cases} \tag{10.21}$$

The Table 10.7 summarizes these results.

If Debreu (1959) thought of his general equilibrium model as a formal abstraction with no idea of being applicable in the "real world" and did not imagine that it could be otherwise, the mathematisation of finance in the 1980s, what we call the second quantification of finance, brought this abstraction into the daily professional

Table 10.7 The "market consistent" convention and the narrative of EMH

Real world probability P		**0.50**	**0.50**
Scenario		Pessimistic	Optimistic
Type of the asset	Asset price at $t = 0$	Forecast for $T = 1$	
A-company share	150	140	200
B-company share	63.89	50	120
Money market risk free rate r	1	1.02	1.02
Arrow-Debreu security (pessimistic)	0.7680	1	0
Arrow-Debreu security (optimistic)	0.2124	0	1
Risk neutral probability Q		**0.7833**	**0.2167**
State price density $L = Q/P$		**1.5666**	**0.4334**
Market pricing kernel $\alpha = L/(1+r)$		**1.5359**	**0.4248**

practice of financiers. Every day, real markets buy and sell Arrow-Debreu prices in transactions based on risk-neutral pricing.

An intersting property of AOA in complete markets is that any income flow from a risky security can be interpreted as a linear combination (weighted average/portfolio) of the Arrow-Debreu securities. Hence, for the risky security A (resp. B) at date $t = 1$, we have the pricing equation:

$$A_0 = A_1(\text{pessimistic}) \times e_0^{\text{pessimistic}} + A_1(\text{optimistic}) \times e_0^{\text{optimistic}}$$

It follows that:

$$\begin{cases} 150.00 = \mathbf{140} \times 0.7680 + \mathbf{200} \times 0.2124 \\ \;\;63.89 = \;\;\mathbf{50} \times 0.7680 + \mathbf{120} \times 0.2124 \end{cases} \tag{10.22}$$

The price system (10.22) reflects the EMH with Arrow-Debreu securities, or EMH-AD.

Thus, the mini-model allows to illuminate the correspondence between the Arrow-Debreu securities, the risk-neutral probability Q and EMH.

10.4.3 The Representation Formats of EMH

In a philosophical paper, Vorms (2011) introduces the notion of representation format as follows.

> I propose to define representation—or representing—as a cognitive activity which consists in using some device (the representing device, or *representans*) in order to gain knowledge concerning something it stands for (the *representatum*).

10.4.3.1 The Alternative Representation Formats

Using the mini-model, it is now possible to easily compare the representations of EMH for the same price system (150.00, 63.89):
EMH-CAPM: First quantification of financial theory—Mean-variance representation

$$\begin{cases} 150.00 = \mathbf{140} \times 0.8824 \times 0.5000 + \mathbf{200} \times 0.8824 \times 0.5000 \\ 63.89 = \mathbf{50} \times 0.7517 \times 0.5000 + \mathbf{120} \times 0.7517 \times 0.5000 \end{cases} \tag{10.23}$$

EMH-Q: Second quantification of financial theory—Risk-Neutral pricing

$$\begin{cases} 150.00 = \mathbf{140} \times 0.9804 \times 0.7833 + \mathbf{200} \times 0.9804 \times 0.2167 \\ 63.89 = \mathbf{50} \times 0.9804 \times 0.7833 + \mathbf{120} \times 0.9804 \times 0.2167 \end{cases} \tag{10.24}$$

EMH-MPK: Market Pricing Kernel representation

$$\begin{cases} 150.00 = \mathbf{140} \times 1.5359 \times 0.5000 + \mathbf{200} \times 0.4248 \times 0.5000 \\ 63.89 = \mathbf{50} \times 1.5359 \times 0.5000 + \mathbf{120} \times 0.4248 \times 0.5000 \end{cases} \tag{10.25}$$

EMH-AD: General Equilibrium theory

$$\begin{cases} 150.00 = \mathbf{140} \times 0.7680 + \mathbf{200} \times 0.2124 \\ 63.89 = \mathbf{50} \times 0.7680 + \mathbf{120} \times 0.2124 \end{cases} \tag{10.26}$$

Each mathematical structure of the model designs a specific representation of EMH, which could be thought as a "representation format" (Vorms 2011).

10.4.3.2 Some Reflections on the Choice of Representations

The use of the simplified mini-model allows one to easily understand how these formats act as fictional representations of the financial "reality", representations that have had a significant importance for professional practices, because of the performativity effect revealed in the impact of financial theories on the business world, meaning that theories shape the real world (MacKenzie 2006). In this sense, representation formats of financial models can be thought as "metaphors" (McGoun 2003) and as "narratives" (Shiller 2019). They are also "explanatory fictions" (Bokulich 2009), fictions that can explain in the sense that they are like helpful fictions that make it possible to design a "useful framework" for practices, namely the "P-world" for portfolio management and risk assessment, and the "Q-world" for asset pricing.

The epistemic contribution of the mini-model is that thus makes it possible to better discover what is fictional in the EMH representations and how the fictions act. The frameworks of real-world probability P and risk-neutral probability Q are two equivalent fictions. To argue this is to support the idea that the probabilistic framework of finance can be considered as a "game with the world" (Shafer & Vovk 2001), a kind of Baudrillard's hyperreality (Macintosh 2003) where "hyperreal" finance means a finance which refers to nothing but itself (McGoun 1997). This financial hyperreality raises the ethical issue of the financialisation of the economy.

Another issue important with the risk-neutral pricing is the performativity of financial models (Callon 1998, MacKenzie 2006). The pricing machinery of SPD "speaks" and what it "says" occurs, like a "speech act" (Austin 1962). Performativity raises the question: what kind of world do we want to see "spoken"? This echoes the ethical issue of financialisation. The financialisation of the economy is a phenomenon that has been both observed and decried for the last thirty years (Epstein 2005). The process of financialisation of the economy has been described in many ways: the growing influence of financial markets in the economic and financial regulation of investments, banking disintermediation, the uncontrolled inventiveness of financial engineering, the growing weight of financial activities in the GDP of developed countries, etc. This process of financialisation brings with it methods of problem analysis, calculation techniques and decision-making principles which, originally created for a limited number of specific cases, have been generalised to all human issues and activities (Chiapello 2015). Among calculation techniques are the risk-neutral pricing methods. Chiapello and Walter (2016) show how the financialisation grew with the conceptual transition from the P-world of the 1960s to the Q-world of the 1980s. A consequence of the shift to the second quantification is that the mathematical machinery makes it possible to price any asset even in the absence of a market for that asset, extending the concept of AOA-"fair value" beyond its original scope. This pricing machinery of financialisation has been named the "financial *Logos*" (Walter 2016). The financial *Logos* is a vector of financialisation.

Let us now consider the topic of green finance under the influence of financialisation. Another key problem for today is the greening of post-crisis finance. Green finance has become a key strategy for the financial industry in the wake of the 2008 financial crisis and understanding green finance becomes increasingly crucial (Jager & Dziwok 2024). Revelli and Walter (2024) have suggested that the risk-neutral pricing could be at the origin of limitations of private green finance strategies, because of the financialisation of these strategies due to risk-neutral paradigm. The Radon-Nikodyn operator could be thought as a "green-destroying operator" that produces natural disasters. In this case, the switch from P to Q with the notion of uniqueness of Q by AOA on complete markets could be seen as a trigger mechanism for financialisation and an obstacle to the greening of finance.

10.5 Conclusion

The narrative of Q risk-neutral pricing has became pervasive in finance. The mini-model can be thought as an conceptual help to understand the presence of this narrative and the narrative itself. It sheds light on the issues related to the valuation of financial assets with the risk-neutral pricing techniques, issues that are actually more conceptual than mathematical. It raises the issue of what does it mean to move from the real world to the risk-neutral world.

The three epistemic gains of the mini-model are the following. It enables "non-quant" financial practitioners and non-specialists of stochastic calculus to understand some of the epistemic difficulties of risk-neutral pricing and, in that sense, it is epistemically valuable and pedagogically useful. It captures one of the most puzzling conceptual aspects of risk-neutral pricing, the uniqueness of the risk-neutral probability Q in a complete arbitrage-free market. It allows one to grasp the signification of the Radon-Nikodyn function L without mathematical expertise, enlightening the interplay between the two mathematical representations (P-world and Q-world) that are used in day-to-day financial practices for financial valuation. At the end, the mini-model paves the way for considering ethical issues related to the pervasive use of the risk-neutral pricing methods.

Acknowledgments This is a new version of a paper prepared for disentengling the notions of fair value, efficient market hypothesis and risk-neutral pricing. Former versions of this paper have been presented in several seminars and conferences in particular the Phinance Biennal Conference, Roma, 8-9 September 2022. I would like to thank Emiliano Ippoliti and participants for their comments. For helpful conversations over the past several years on topics covered in this paper, I want to thank Jean-Philippe Bouchaud, Ernst Eberlein, Monique Jeanblanc, Olivier Le Courtois and Dilip Madan. I am grateful to Robert Dimand for helpful comments on the history of uncertainty in economics and to Ewa Dziwok for many constructive remarks and for checking the mathematics of this paper. I thank two anonymous referees for helpful comments on its structure and content. I also benefited of conversations with practitioners as Hubert Rodarie who provided relevant insights on european financial regulation. All errors remain mine.

References

Arrow, K. J. (1959). Rational choice functions and orderings. *Economica, 26*(102), 121–127. https://doi.org/10.2307/2550390

Arrow, K. J., & Debreu, G. (1954). Existence of an equilibrium for a competitive economy. *Econometrica, 22*(3), 265–290. https://doi.org/10.2307/1907353

Austin, J. L. (1962). *How to do things with words*. Oxford University Press.

Black, F., & Scholes, M. (1973). The pricing of options and corporate liabilities. *Journal of Political Economy, 81*(3), 637–654. http://www.jstor.org/stable/1831029

Bokulich, A. (2009). Explanatory fictions. In M. Suarez (Ed.), *Fictions in science: Philosophical essays on modeling and idealization*. Routledge.

Boyle, P., & Boyle, F. (2001). *Derivarives. The tools that changed finance*. Risk Books.

Callon, M. (1998). What does it mean to say that economics is performative? In D. MacKenzie, F. Muniesa, & L. Siu (Eds.), *Do economists make markets?* (pp. 310–357). Princeton University Press.

Cartwright, N. (2009). If no capacities then no credible worlds. *Erkenntnis, 70*, 45–58.

Chiapello, E. (2015). Financialisation of valuation. *Human Studies, 38*(1), 13–35.

Chiapello, E., & Walter, C. (2016). The three ages of financial quantification: A conventionalist approach to the financier's metrology. *Historical Social Research, 41*(2), 155–177.

Cochrane, J. (2001). *Asset pricing*. Princeton University Press.

Cont, R. (2001). Empirical properties of asset returns: Stylized facts and statistical issues. *Quantitative Finance, 1*(2), 223–236.
Cox, J., Ross, S. A., & Rubinstein, M. (1979). Option pricing: A simplified approach. *Journal of Financial Economics, 7*(3), 229–263.
Desrosières, A. (2003). Managing the economy: The state, the market and statistics. In T. M. Porter & D. Ross (Eds.), *The Cambridge history of science. Modern social and behavioral sciences* (vol. 7, pp. 553–64). Cambridge University Press.
Diebold, F. X., Doherty, N. A., & Herring, R. J. (2010). *The known, the unknown, and the unknowable in financial risk management*. Princeton University Press.
Dimand, R. (2021). Keynes, Knight, and fundamental uncertainty: A double centenary 1921–2021. *Review of Political Economy, 33*(4), 570–584.
Dothan, M. (1990). *Prices in financial markets*. Oxford University Press.
Duffie, D. (1996). *Dynamic asset pricing theory* (3rd edn.) [2016]. Princeton University Press.
Ellsberg, D. (1961). Risk, ambiguity, and the savage axioms. *The Quarterly Journal of Economics, 75*(4), 643–669.
Epstein, G. (2005). *Financialization and the world economy*. Edward Elgar Publishing.
Fama, E. (1970). Efficient capital markets: A review of theory and empirical work. *Journal of Finance, 25*(2), 383–417 and discussion pp. 418–423.
Fama, E. F. (1991). Efficient Capital Markets: II. *The Journal of Finance, 46*(5), 1575–1617. https://doi.org/10.2307/2328565
Föllmer H., & Schied, A. (2002). *Stochastic finance. An introduction in discrete time*. de Gruyter.
Gilboa, I., & Schmeidler, D. (2012). *Case-based predictions: An axiomatic approach to predictions, classification and statistical learning*. World Scientific Publishing.
Habart-Corlosquet, M., Gehin, W., Janssen, J., & Manca, R. (2015). *Asset and liability management for banks and insurance companies*. ISTE Wiley.
Habart-Corlosquet, M., Janssen, J., & Manca, R. (2013). *VaR methodology for non-Gaussian finance*. ISTE Wiley.
Harrison, J. M., & Kreps, D. M. (1979). Martingales and arbitrage in multiperiod secutities markets. *Journal of Economic Theory, 20*(3), 381–408.
Harrison, J. M., & Pliska, S. R. (1981). Martingales and stochastic integrals in the theory of continuous trading. *Stochastic Processes and Applications, 11*(3), 215–260.
Hempel, C. (1966). *Philosophy of natural science*. Prentice-Hall.
Ippoloti, E. (2017). Methods and finance. A view from inside. In E. Ippoliti & P. Chen (Eds.), *Methods and finance. A unifying view on finance, mathematics and philosophy* (pp. 121–128). Springer.
Jager, J., & Dziwok, E. (Eds.). (2024). *Understanding green finance. A critical assessment and alternative perspectives*. Edward Elgar Publishing.
Jarrow, R., & Larson, M. (2012). The meaning of market efficiency. *Mathematical Finance, 22*(1), 1–30.
Jondeau, E., & Rockinger, M. (2006). Optimal portfolio allocation under higher moments. *European Financial Management, 12*(1), 29–55.
Jurczenko, E., & Maillet, B. (2012). The four-moment capital asset pricing model. Between asset pricing and asset allocation. In E. Jurczenko & B. Maillet (Eds.), *Multi-moment asset allocation and pricing models* (pp. 113–163). Wiley.
Kahneman, D., & Tversky, A. (1979). Prospect theory: An analysis of decision under risk. *Econometrica, 47*(2), 263–291.
Karatzas, I., & Shreve, S. E. (1998). *Methods of mathematical finance*. Springer.
Kemp, M. (2009). *Market consistency. Model calibration in imperfect markets*. Wiley.
Knight, F. H. (1921). *Risk, uncertainty and profit*. Houghton Mifflin Company.
Le Courtois, O., & Walter, C. (2016). *Extreme financial risk and asset allocation*. Imperial College Press.
Le Courtois, O., & Xia, X. (2023). Efficient portfolios and extreme risks: A Pareto-Dirichlet approach. *Annals of Operations Research, 335*, 261–292. https://doi.org/10.1007/s10479-023-05507-y

Lucas, R. E. (1978). Asset prices in an exchange economy. *Econometrica, 46*(6), 1429–1445.
Macintosh, N. (2003). From rationality to hyperreality: Paradigm poker. *International Review of Financial Analysis, 12*, 453–465.
MacKenzie, D. (2006). *An engine, not a camera. How financial models shape markets*. The MIT Press.
Markowitz, H. (1952). Portfolio selection. *The Journal of Finance, 7*(1), 77–91.
McGoun, E. G. (1997). Hyperreal finance. *Critical Perspectives on Accounting, 8*(1–2), 97–122.
McGoun, E. G. (2003). Finance models as metaphors. *International Review of Financial Analysis, 12*(4), 421–433.
McNeil, A., Rudiger, F., & Embrechts, P. (2015). *Quantitative risk management. Concept, techniques and tools* (revised version). Princeton University Press.
Ottaviani, G. (Ed.). (1995). *Financial risk in insurance*. Springer.
Pickering, A. (1992). *Science as practice and culture*. The University of Chicago Press.
Pliska, S. (1997). *Introduction to mathematical finance. Discrete time models*. Blackwell.
Reghai, A. (2015). *Quantitative finance. Back to basic principles*. Palgrave Macmillan.
Reutlinger, A., Hangleiter, D., & Hartmann, S. (2018). Understanding (with) toy models. *The British Journal for the Philosophy of Science, 69*(4), 1069–1099.
Revelli, C., & Walter, C. (2024). Limitations of conventional private green finance industry and strategies. In J. Jager & E. Dziwok (Eds.), *Understanding green finance. A critical assessment and alternative perspectives*. Edward Elgar Publishing.
Savage, J. (1954). *The foundations of statistics*. Wiley.
Schelling, T. (1971). Dynamic models of segregation. *The Journal of Mathematical Sociology, 1*(2), 143–186.
Shafer, G., & Vovk, V. (2001). *Probability and finance. It's only a game!* Wiley.
Sharpe, W. (1964). Capital asset prices: A theory of market equilibrium under conditions of risk. *The Journal of Finance, 19*(3), 425–442.
Shiller, R. (2019). *Narrative economics: How stories go viral and drive major economic events*. Princeton University Press.
Svetlova, E. (2021). On the relevance of Knight, Keynes and Shackle for unawareness research. *Cambridge Journal of Economics, 45*, 989–1007.
Svetlova, E., & Dirksen, V. (2014). Models at work – models in decision making. *Science in Context, 27*(4), 561–577.
Vorms, M. (2011). Representing with imaginary models: Formats matter. *Studies in History and Philosophy of Science Part A, 42*(2), 287–295.
Walter, C. (1996). Une histoire du concept d'efficience sur les marchés financiers. *Annales. Histoire Sciences Sociales, 51*(4), 873–905.
Walter, C. (2001). The efficient market hypothesis, the Gaussian assumption, and the investment management industry. EFMA 2001 Lugano Meetings, SSRN: https://ssrn.com/abstract=267443
Walter, C. (2016). The financial Logos: The framing of financial decision-making by mathematical modelling. *Research in International Business and Finance, 37*, 597–604.
Walter, C. (2017). Philosophie de la finance : l'exemple de l'efficacité informationnelle d'un marché. In G. Campagnolo & J.-S. Gharbi (Eds.), *Philosophie économique. Un état des lieux* (pp. 579–626). Éditions matériologiques.
Walter, C. (2019). The Brownian motion in finance: An epistemological puzzle. *Topoi, 40*, 1–17.
Wütrich M. V., Bühlman, H., & Furrer, H. (2015). *Market-consistent actuarial valuation*. Springer.
Ylikoski, P., & Aydinonat, E. (2014). Understanding with theoretical models. *Journal of Economic Methodology, 21*(1), 19–36.

Chapter 11
The New Form of Digital Money: Not a Mere Technological Innovation

Fabrizio Zennaro

Abstract The evolution of new forms of digital money has been going on for over a decade. One of the enabling elements of this development is technological innovation. However, first applications of crypto-assets do not represent digital currencies, but an evolutionary step towards monetary instruments. To understand the phenomenon and imagine future scenarios, it is necessary to analyse the functioning of money from a technical-monetary perspective. With these tools, it is possible to frame better the contribution of technological innovation, both for private and public forms of digital money. Outlining the contribution in this way, it emerges that technology is not the only innovation factor to consider. Market competition and regulatory responses play also a driving role, combined and interconnected with technology. Following this approach, the future of digital currencies should be envisaged bearing in mind also the relevance of the demand for money in terms of needs as well as the direct and reflexive effects of regulation.

11.1 Introduction

The new millennium has brought a strong acceleration of technological innovation. Among the various sectors, financial and payment industries experienced one of the highest rates of innovation. On these aspects there are extensive debates on the possible evolution of money and on the impacts deriving from new technologies. These analyses usually aim to identify future scenarios or to outline guidelines for

---

The opinions expressed in this article are those of the author and do not necessarily reflect the official policy or position of the Bank of Italy.

F. Zennaro (✉)
Banca d'Italia, Rome, Italy

Sapienza University, Rome, Italy
e-mail: fabrizio.zennaro@uniroma1.it

E. Ippoliti et al. (eds.), *Philosophy and Finance: Ten Open Questions*, Synthese Library 504, https://doi.org/10.1007/978-3-031-96510-4_11

the decisions of public agents. In many of these publications, technology is framed as a facilitator of innovation that creates new markets.

If it is certainly true that technological innovation allows the offer of new services, nevertheless it is not the only driving force behind new forms of digital money. There are other two directly interrelated: the competitive factor and the regulatory context. The co-presence of these generates the reference context on which technological innovation starts and develops. In that sense, new forms of digital money are not just a mere technological issue.

According to this perspective, additional factors influence innovation and therefore the emerging of new forms of money. Market demand stimulates the competitive initiative on the supply side, while the reflexive effects of regulatory frameworks generate its unpredictable development.

The proposed analysis crosses different disciplines, and—from their relationship—I represent monetary innovation from a broader perspective, aiming at offering an additional tool both for private and public agents. In doing so I will mainly use the regulatory and monetary references present in the European Union, while for crypto-assets I will use the terminology widespread in the market.

The document is structured as follow. In the second paragraph, I identify the essential elements of monetary economics to clarify what digital money is and how it relates to the financial market and the payment system, defining four useful elements to contextualize the technological impact. In the third and fourth paragraphs I focus the analysis on technological innovations related to the development of emerging digital currencies, both from private and public initiatives. In the fifth paragraph, I identify the driving forces that are directly and indirectly interconnected with technology. In the conclusions, I present the future perspectives, evaluating the regulatory initiatives and the possible responses from the market, according to the analysis conducted.

11.2 Understanding What Digital Money Really Is

Money is anything that can serve as store of value, unit of account and medium of exchange (IMF, 2012).[1] For its functioning, it has to be simple to use. However, behind an opportunity to pay with a piece of paper depicting a bridge or a transaction confirmed with a sound emitted by a watch, there is a complex system that determines money acceptability among people. When dealing with the issue of digital money, one must keep in mind a relevant but not obvious distinction: the difference between good and bad money, in terms of adherence to the fulfillment of the three traditional functions. Especially for advanced economies, only good

[1] The store of value allows people to save and use it later, smoothing their purchases over time; the unit of account provide a common base for prices; the medium of exchange helps people to buy and sell from one another (IMF, 2012).

money can be a possibility, otherwise there would be economic and financial degradation. In other words, if anything can be used as a money instrument, not everything is compatible with modern economic systems.

Considering that I want to analyze the prospects of a good digital money, it is necessary to evaluate its characteristics as well as the process underlying its functioning. Some fundamental concepts of monetary policy are complex and could lead non-technicians to fallacious reasoning resulting from incorrect assumptions. There are at least four interconnected concepts to be taken in consideration: the economic functions of money, the role of trust, the relationship between public and private money, the creation of money in the modern monetary system.

The first is the simplest and most intuitive one, relating to the three economic functions of money commonly considered: medium of exchange, store of value and unit of account. These three simple functions primarily facilitate the exchange of goods and services, but consequently also the economic development of a society. The development of financial services, and in particular of credit, is closely connected to the function of store of value and to the creation of money. Thus it is essential to have the availability of a currency in which all three functions are fulfilled. Looking ahead, we must evaluate the evolution of digital money that is able to fulfill these functions, thus consisting in "good digital money". This concept is consistent with the vision of the social function of money, where an object (even digital) is money if it objectively performs the three functions (Guala, 2016). The necessary assumption is that money has to perform well in modern economies.

The second element is related to the role of trust in money. Over time, money has evolved in the way it represents value. It has gone from being an object (coins) containing precious metal, later to banknotes convertible into gold, up to the current status where its value is not intrinsic, but it is legally defined by State power (fiat money). The trust is empowered by the confidence in the ability of the central bank to keep its value stable, and it is indeed the founding element for the functioning of the entire monetary system. It is not achieved through a simple association to the State trustiness but, on the contrary, through a complex process that involves the supply and demand for money. In modern economies the management and control of money has been made independent from governmental powers, delegating by law this task to specific authorities (e.g. Central banks), with binding objectives and financial autonomy to achieve them.

In the current state of development, therefore, trust in the currency is "gold". In particular, this trust is expressed in the process of issuance, transfer and conservation (De Bonis & Ferrero, 2022a, 2022b): the trust that money can be accepted in the future at a substantially stable value, the trust of freeing oneself from obligations when it is accepted,[2] the confidence that the instrument used is actually money. Technological evolution, as we will see in the next paragraph, aims to provide

[2] Banknotes and coins are guaranteed by the "legal tender" status, that foresees (1) an obligation to accept and the power to extinguish the payment obligation and (2) the acceptance at full face value.

automated solutions to feed trust, aimed at disintermediating the role of central and commercial banks.

The third element, the type of money supply, contains technicalities that take the discussion out of the common debate and, above all, increase the complexity that makes it difficult to formulate linear conclusions. I am referring to the relation between public and private money.[3] In modern economies, we experience a coexistence between public and private money, the so-called "two-tiered system". Public money is issued by the Central bank, while private money by commercial banks or electronic money institutions.

In case of public money, Central bank issues two forms of money that constitute the "base money": (1) "retail money", currency (banknotes and coins)[4] used by all economic agents as citizen and firms and (2) "wholesale money", central bank reserves, available only to banks to carry out transactions among themselves or with the central bank.

In case of private money, commercial banks issue their own private money, called "bank deposits", under the supervision of the Central bank. In addition, electronic money institutions, in addition of provisioning of payment services, can issue, redeem and distribute electronic money, under specific limitations and supervision. The different forms of public and private money are expressed in the same unit of account and thus have a conversion rate at par at all times. The representation described is a simplification of the whole process of the monetary system,[5] but it is sufficient to represent three fundamental concepts: (1) the money we use for retail payments is not all public, indeed the latter represents a minority;[6] (2) a digital public currency already exists, i.e. central bank deposits, necessary for the functioning of the modern centralized payment system;[7] (3) there is already a digital money available to retailers, i.e. bank deposits and electronic money, which enable digital payments (Fig. 11.1). The current money supply is therefore much more complex than what appears behind the common erroneous idea of money with a single type of issuance.

[3] Some authors refer to inside or outside money, where the terms refer to the fact that money is backed by debt/assets from inside/outside the private sector.

[4] In the monetary union, only the European Central Bank (ECB) and National Central Banks (NCBs) of the participating Member States, subject to authorization by the ECB, are empowered to issue euro banknotes. Member States may issue euro coins, subject to approval of its volume by the ECB.

[5] For a detailed representation, see "Moneta pubblica e privata in un sistema economico moderno" (Ferrari & Ferrero, 2020), where the implications are effectively described through the representation of stylized balance sheets.

[6] In Italy, where the currency circulation is high, banknotes amount to 259 billion euros, while bank current account deposits are 1452 billion euros (data as of May 2023, source Bank of Italy).

[7] In retail payments is crucial the trust in the convertibility from private to public money. Commonly used tools such as ATMs convert private money into public money, while credit cards or bank transfers convert private money from various bank issuers.

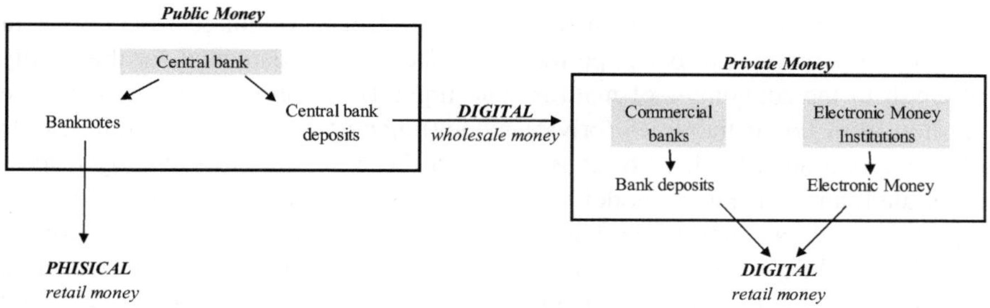

Fig. 11.1 The (simplified) two-tiered monetary system

The fourth element is the most complex one, interconnected with the elements previously described. I am referring to the process of money creation. Again, the most widespread image is the simplified one of the central bank printing banknotes. In reality, the process of monetary creation is concentrated more in digital form, both in the context of public and private money. The process of creating public money takes place through the printing of additional banknotes or, digitally, through lending operations to banks[8] or through the purchase of financial assets.[9] These operations, by adding the monetary base available in the system, vary the amount of reserves available to the banking system.[10] Similarly, private money is also created by commercial banks through the granting of loans (to households and businesses) and the purchase of financial assets from the private sector. Private money creation comes with risks[11] (monitored by supervision authorities) and has limitations, resulting from the demand and supply of money.[12] Since private money is the most widespread for the purchase of goods and services, this is the relevant type for controlling price stability through Central banks' monetary policy operations.

To fully understand the process, we must also consider that commercial banks in advanced economies operate in a "fractional-reserve system", a system in which commercial banks are allowed to grant credit to citizens and businesses by holding

[8] Refinancing operations. In this context, the Central bank plays the role of bank of banks, lending public money against collateral which is used for transactions between private institutions or for maintaining the required reserve (see note 13).

[9] The outright asset purchase.

[10] To these are added some variations—called autonomous factors—which do not depend on the management of the Central bank. For example, the demand for currency and variations in the government account at the Central bank.

[11] Uncontrolled issuance carries the risk of price instability, while excessive lending exposes to the risk of financial instability.

[12] For example, limits on the demand for new money for payment purposes is influenced by interest rates and limits on the supply from the constraints imposed by prudential regulation on the composition and size of balance sheets. For more details, see Ferrari and Ferrero (2020).

only a corresponding fraction in reserves Central bank (required reserves[13]). In this system, commercial banks perform the function—fundamental for the supply of credit to the economy—of maturity and liquidity transformation, i.e. the issue of short-term liquid liabilities (private money) against longer-term illiquid assets (the credits granted). This aspect is important because works differently respect to an alternative operating model, called the "narrow system", in which a private liquid liability can only be granted against a full match of liquid asset (usually issued by public sector). This system, recalled during the financial crises as a solution to systemic risks, effectively nullifies the distinction between public and private money, modifying the current monetary functioning system. It is considered inefficient, as it would reduce excessively the lending capacity.[14]

Let me recap. These four fundamental concepts highlight the complexity of the monetary system and the interconnections between public control and private initiative. Above all there is trust in the money, necessary to ensure certainty in exchanges. Trust is based, among other things, on the functioning of payment systems and on price stability over time. Furthermore, the economy is based on the provision of credit by banks. The latter are responsible, in a competitive market, for risk assessment and maturity transformation. The joint and interconnected presence of public and private money guarantees the system an efficient mechanism for allocating credit.

The digital money currently available, Central bank deposits, commercial bank deposits and e-money, have the characteristics to be considered "good digital money". The evolution of digital currency, enabled by the new technology, will necessarily develop in accordance with the three functions but also with the other functional elements of the monetary system.

11.3 The Technological Innovation in the Private Money

Technology certainly represents a driving force for innovation and the pursuit of digital currency is not a recent phenomenon, but rather the result of decades of experimentation (see Brunton, 2019). It is therefore important to evaluate its impact using the four monetary elements identified, tackling separately private and public money. I will begin by discussing the technological evolution that first involved private money and then, in the following paragraph, I will describe the options for the development of public money.

---

[13] The required reserve started as a prudential tool to contain the liquidity risk in the event of a bank run. It became a monetary policy tool to control inflation, as the change in the official rates applied to the reserves at the Central Bank are transmitted indirectly to the interbank market, thereby influencing the amount of private money created in the market.

[14] In the narrow-system loans would be granted only against the ability to issue of securities of similar maturity as the credit granted, or deferred to the Central Bank's willingness to offer reserves and thus directly manage the transformation of maturities.

The spread of Distributed Ledger Technologies[15] (DLT), in the more specific form of the blockchain,[16] has represented the innovative basis for the development of the market of so-called cryptocurrencies, more correctly defined as crypto-assets.[17] Using the market terminology, two stages of technological evolution of crypto-asset can be identified: Bitcoin and Stablecoins.

The first can be considered as the origin of the phenomenon in early 2009 as a response to the financial crisis, after the white paper on the Bitcoin crypto-asset was published with the aim of creating a currency without the presence of centralized intermediaries. The great technological innovation is the validation system of transactions based on a distributed register (see also Auer et al., 2023), defined as blockchain, which has spread rapidly in various sectors also affecting the provision of traditional services, such as trading platforms and payment services. However, Bitcoin has not represented a concrete threat to the current monetary system, both due to its intrinsic volatility which makes it unsuitable to be used as a payment instrument.[18]

In terms of the three monetary functions, Bitcoin is not able to simultaneously perform the three functions that characterize money: it is a unit of account, a (risky) reserve of value, but it does not have the liquidity characteristics that would allow it to be effectively used as a means of payment. From a legal perspective some authors argue that Bitcoin, as anything that calls itself money, is a currency if in fact it performs the functions that characterize it as such in a given community (Moliterni in Maimeri & Mancini, 2019). The quality of the characteristics would therefore not be important (whether it is good money) but it is a question of "currency", or if in a specific community the qualities or functions are recognized (means of exchange and store of value).[19] In this sense, in a definition based on technological neutrality, cryptocurrencies and private digital currencies end up coinciding, where the essentiality of cryptography represents the "ontological" characteristics of virtual currencies (*ibidem*). Anyway, from a monetary point of view, the function as a medium of exchange is not performed as it should be and so Bitcoin cannot strictly be considered money. The result is that—as a means of payment—it is used for illicit activities such as money laundering, financing of terrorism and tax evasion (BIS, 2023).

---

[15] It consists of a shared electronic ledger with data protected both by cryptographic techniques and by means of 'redundancy' (Banca d'Italia, 2022).

[16] Specifically, it is called blockchain because the memorized transactions are grouped in a sequence of 'blocks' linked cryptographically, creating a chronological record that cannot be altered of all transactions made up to that point in time (Banca d'Italia, 2022).

[17] Markets in Crypto-Asset Regulation (MiCAR), art. 3 (5) define crypto-asset as "a digital representation of a value or of a right that is able to be transferred and stored electronically using distributed ledger technology or similar technology".

[18] The characteristics of Bitcoin that we will analyze below are common to many other crypto-assets, but we can refer to Bitcoin as it represents the largest market capitalization.

[19] Extract from the text by Francesco Moliterni, "*Criptovaluta, valuta digitale, moneta elettronica e modelli di circolazione*" (Maimeri & Mancini, 2019), pp. 190 and 191.

The element of trust is the most interesting topic, as the objective of blockchain technology is to disintermediate the current intermediaries (Central banks and commercial banks) from the issuance, transmission and storing process. Disintermediation is pursued through distributed ledger technology and implemented through cryptographic algorithms, which allow transactions to be validated through the consensus mechanism[20] as well as to verify transactions in an open and verifiable mode for all. Bitcoin therefore represents the crypto-asset, while the blockchain embodies the infrastructure for its circulation and storing. The functioning of the consensus mechanism is also an essential element for trust in the currency, where for example in Bitcoin the current proof-of-work mechanism does not represent a perfect and unshakable system, especially in the absence of an institution that oversees the overall process.

The third monetary element concerns the type of supply. Bitcoin is characterized by the absence of forms of user-protection, unlike what is provided for private money issued by supervised entities (as commercial banks). Deposit insurance, supervision of authorized issuers, the value of legal tender and, above all, convertibility at par with public money, constitute essential elements for monetary anchoring and therefore trust in private money. All these features are absent in cryptographic forms of assets as Bitcoin.

Finally, with regard to the issuance element, it is important to note that the creation of Bitcoins is pre-established and its distribution take place as a remuneration for the PoW validation process;[21] it is therefore linked to an algorithm rule rather than a demand factor. Furthermore, the same computer code also foresees a limit on creation, set upon reaching the total supply of around 21 million Bitcoins. When this threshold will be met, new Bitcoins will no longer be created by the mining process and the validation of transactions will be remunerated exclusively for commissions. This rule implies that in the long run Bitcoin, if used for payment purposes, will become a "deflationary" crypto-asset like gold in the past[22] (De Bonis & Ferrero, 2022a, 2022b).

Through the analysis of the four monetary elements it is therefore possible to define that Bitcoin is not a digital currency, but rather a highly speculative investment asset.

The second phase of technological innovation refers to the evolution of a second generation of crypto-assets, called stablecoins,[23] which has led to a growing level of attention from Central banks and public institutions. The innovation is

[20] For Bitcoin is the so called "proof-of-work" (PoW), but there are many others that try to overcome specific limitations.

[21] The mining algorithm foreseen that the remuneration of the PoW through Bitcoin should halve around every 4 years.

[22] An increase in demand relative to supply corresponds to an increase in the price of money. If used as a means of payment, it leads to a downward alignment in the prices of goods and services (deflation).

[23] A digital asset that attempts to anchor its value to a reference price, in order to ensure its stability through underlying assets.

here represented by the price stabilization mechanisms that potentially allow the volatility limit to be overcome, favoring its use as a means of payment. In particular, the public presentation of the first white paper of the stablecoin Libra in June 2019 generated the reaction of the main institutions, specifically for the potential impacts as financial stability of the so called "global stablecoin", and led to a rapid publication of the report of the G7 working group on Stablecoins.[24]

When analyzing stablecoins through the lens of the four monetary elements, we must keep in mind that the term "stablecoin" contains a universe of crypto-assets, which differ from each other according to the technical characteristics of the stabilization mechanism and other relevant aspects as the issuer governance (see for example Bullmann et al., 2019).

Regarding the first element, the functions of the currency, the satisfaction of the condition of "good digital money" depends on the type of stablecoin. There is the case when could be consistent with that of existing money, when the stabilization mechanism provides for the possibility that the price of the stablecoin is linked to a fiat currency and the value guaranteed by corresponding assets (asset-linked stablecoin), as the specific type of stablecoins classified be the European legislation as "electronic money token" (EMT), under specific rules for their issuance and supervision.[25] In the other cases where the stabilization mechanism does not fully guarantee full 1 to 1 convertibility with a fiat currency, however, the quality of the three money functions may be less degraded.

The stabilization mechanism has certainly provided a step forward in trust, particularly regarding the ability to maintain value over time. However, the quality of this mechanism determines how close this level of trust can be to that of current digital fiat currencies. In some extreme case, as the crypto-assets defined as "algorithmic stablecoins", the price stabilization mechanism is linked to an algorithm that determines the supply based on demand with no value guarantee funds (unbacked-stablecoins). These mechanisms are far from configuring trust, as experienced in the case of the Terra-Luna algorithmic stablecoin in 2022, where due to a presumed premeditated attack, the value of the algorithmic stablecoin disengaged from the dollar and subsequently plunged dramatically.

As the type of money supply is private, the guarantees for users are relevant and depend on the presence of regulation. For example, where regulation is present there will be more guarantees, whereas in their absence these could not be foreseen.

Regarding money creation, the possibility that stablecoins are issued 1 to 1 compared to a fiat currency as collateral would make them similar functioning to the existing electronic money, therefore issuing as in a narrow-system model. Otherwise, if funds were less, the system would be more similar to the fractional reserve model. In this case, there would be closer relation with the commercial banks account, the latter under the supervision of the issuer's operations.

---

[24] BIS (2019).

[25] The regulation also foresees another distinct form of stablecoins, the asset-reference-tokens (ARTs). These crypto-assets refer the price to a basket of currencies or assets to which correspond defined guarantees.

In conclusion, through the analysis of the four monetary elements it is possible to imagine that some types of stablecoins, in particular some regulated ones with characteristics similar to e-money, could take on the use of future digital currencies. However, the possibility that specific stablecoins will actually spread as a payment instrument will also depend on other driving forces, which I will analyze in the following paragraphs.

11.4 The Technological Innovation in the Public Money

Let us now focus on technological innovation of the potential new digital public currency, namely the Central Bank Digital Currency (CBDC), both in the retail and wholesale type. The current debate on the possibility of issuing a new public money, expanding the supply of monetary base, aims at adapting to a changing context. To better understand the different options, it is necessary to make a premise on the context.

In the last decade, two particularly relevant factors have occurred: (1) for the demand, the evolution towards a cashless society with the reduction in the demand for cash and (2) for the supply, the appearance of potential new forms of digital money that cannot be controlled by the monetary authorities, both in private and public form, as the case respectively of stablecoins and foreign CBDCs[26] (Fig. 11.2).

On the demand side, the first scenario concerns the customer needs for a reduction on the use of banknotes, that could drive the monetary system towards a "cashless society". For example, in the European Union the demand for banknotes is declining, albeit at different speeds within countries. Sweden, in particular, is among the countries experiencing a very low use of cash (Sveriges Riksbank, 2017).

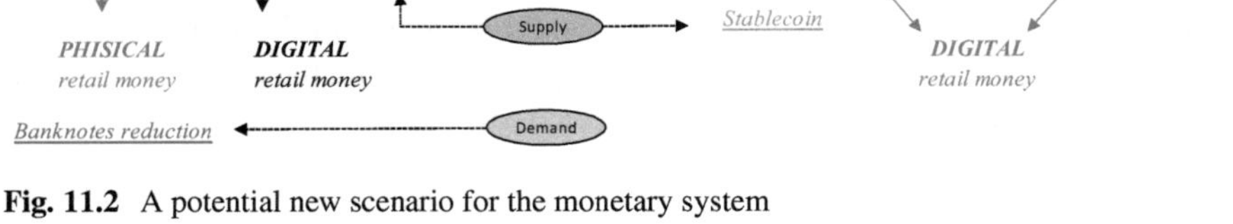

Fig. 11.2 A potential new scenario for the monetary system

[26] A foreign CBDC is a CBDC issued by authorities of other economic areas.

For this reason, the Riksbank, the Swedish Central bank, was among the first to investigate the possibility of issuing a CBDC.[27] One of the problems highlighted was linked to circulation; if in Sweden cash were no longer used for transactions in shops in favor of credit cards, it would mean, among other things, that the payment system relies exclusively, for the point of sale, on the private infrastructure of the card issuers, which are mainly foreign private operators (*ibidem*), guided by the logic of profit rather than continuity of the public service. Even in countries where the use of cash is still consistent, such as Italy, the decline is evident. The increase in the use of digital currency for transactional purposes is accompanied by multiple phenomena.[28]

On the supply side, a novelty is the potential emergence of the stablecoins. As mentioned, the prospect of having a "competitive" private money not controlled by the monetary authorities emerged in 2017, when Facebook announced the Libra stablecoin project. The great concern was linked to the relevance of the Facebook network, which at the time counted around three billion accounts, which could create the condition for a potential a new unit of account. The risk identified by the authorities was therefore the so-called "global stablecoins". Although the Libra (renamed Diem) risk wound down with the abandonment of the project,[29] other stablecoins will soon be a reality. While a specific regulation is not yet envisaged in the United States, in the European Union the aforementioned MiCAR will apply from 2024, thus giving the opportunity of developing regulated stablecoins. The European initiative will impact on a wider range than its regional perimeter; recently, the payment company Paypal and the blockchain provider Paxos launched the "Paypal stablecoin", probably the first potential global stablecoin entering in the regulatory perimeter (of MiCAR).

Another incumbent evolution in the supply is represented by the possibility that foreign countries make their own CBDCs available on the market, thus creating an alternative to national currencies (foreign CBDCs). Although the Federal Reserve, the European Central Bank, or the Bank of England are still in the evaluation or investigation phase,[30] the Bank of China[31] and the Bank of India[32] are already in a pilot phase, which should precede the launch of a CBDC. The competition between

---

[27] The first CBDC publication of the "e-Krona project" was in 2017.

[28] For example, a study by the Bank of Italy on the use of cash under the 2019 pandemic has highlighted the increase in the use of electronic transactions, albeit accompanied by an increase in cash withdrawals (Ardizzi et al., 2020). This apparently contradictory result reveals that purchasing habits had shifted towards the use of digital money, while banknote withdrawals were mainly aimed at the purpose of store of value (Bank of Italy, 2023).

[29] After the launch of the initiative, some important players such as PayPal, eBay, Visa and Mastercard abandoned the project also due to the pressure and concerns of regulators.

[30] For the international state of progress see https://cbdctracker.org/.

[31] For more information, see Jan Knoerich *China's New Digital Currency: Implications for Renminbi Internationalization and the US Dollar* (Bilotta & Botti, 2021).

[32] India has developed a digital financial architecture (UPI or Unified Payment Interface) which already provides for integration with a probable retail CBDC.

currencies, in addition to the strength of its functions, could also play a role on the strategy for the competitive advantage of first mover.

From the scenario described it is evident that the current monetary system could be subjected to imbalances that are difficult to predict in terms of effects and magnitude. The monetary authorities have expressed themselves very clearly in this regard. The ECB recalled the importance of the monetary anchor, the coexistence of public and private money in the "two-tiered system" (Panetta, 2022). The monetary anchor guarantees trust in private money deriving from the possibility of always converting it at par with public money (banknotes). If, as described, the use of banknotes almost disappears or uncontrolled private currencies spread, the two-tiered system would be mined, with implications for the credit function offered by financial intermediaries. The "CBDC retail" project fits into this context, representing a digital currency, public and accessible to retail users, in the form of a current account at the Central bank or a digital banknote (see Auer & Bohme, 2020).

It is interesting to note that the development of retail CBDC is not necessarily linked to technological innovation. In the investigation phase promoted by the ECB and started in 2020[33] it is clearly stated that the DLT solution is not the basis for the choice of a possible issuance, while other characteristics—more directly linked to money—become fundamental, such as the management of any interests, the status of legal tender, privacy control and the method of authentication. The change of these characteristics leads to significant impacts on the payment system, the conduct of monetary policy, financial stability and banking functioning. For example, the technical solution on the verification process for payments configures two different solutions: in the account-based system it is verified the identity of the payer, thus the CBDC is configured similarly to a bank account; in the token-based system it is verified the authenticity of the exchanged object, so CBDC is more in analogy with banknotes. The technical solution, even before the technological one, defines the overall scenario of a CBDC for the efficiency of the retail payment system, the control of crimes, the strengthening of monetary policy as well as the impacts on monetary sovereignty. To give an idea, among different hypotheses, there is an account-based CBDC to which a two-tiered remuneration system is associated. This solution would configure a CBDC as a Central bank deposit open to retail, with a double interest rate applied on different amounts. This would allow demand of public money to be controlled in a similar way to what already happens for Central bank reserves (Bindseil, 2019, 2020). I leave the details to the numerous and specific insights, but I would like to highlight once again how these aspects are more inherent to the processes rather than to specific technologies.

The decision regarding the issuance of a CBDC is a highly strategic choice that may differ from country to country. In the debate on the opportunity of a digital euro, for example, the risk of concentration on technology companies, mostly foreign, is very relevant, in particular the risks of concentration on the overall digital money offer of dominant positions through technological platforms, issues relating

[33] See "Report on the digital euro" (ECB, 2020).

to privacy and the correct functioning of the payment system, monetary sovereignty and financial stability (Panetta, 2023).

In addition to the retail CBDC above described, a further potential new public currency is being evaluated: the "wholesale CBDC", a digital currency intended for exchanges between authorized financial intermediaries. In this case, as mentioned, there is already a public digital currency for wholesale exchanges. In the Eurosystem, the central bank deposits are used to settle exchanges between commercial banks through TARGET2 and TIPS systems. The elements of functionality, trust, public nature and creation would remain almost unchanged. Also in this case, the technological element is only potentially a pivot of innovation. The Banque de France, for example, proposes a payment system between authorized intermediaries based on DLT technology, such as to allow the programmability function of transactions and to follow the emerging trend of tokenisation of finance.[34] But, even in this case, the technological element is a potential enabler and not necessarily a factor of innovation. In fact, as an operational counter-proposal, the Bundesbank and the Bank of Italy presented projects, respectively for Target2 and TIPS,[35] which would ensure the same programmability functionality by adding a specific protocol to existing systems (Bank of Italy, 2021). The decision to develop a wholesale CBDC, also in this case, goes beyond mere technological innovation.

The development of new digital currencies, public and private, are not only or necessarily linked to innovative technologies such as DLT and Blockchain. These have opened up new potential services, but the driving forces of innovation also depend on the presence of additional contextual factors, analyzed in the following paragraph.

11.5 Driving Forces for New Forms of Digital Money

Technology has allowed the development of new services. However, the ecosystem is complex and other equally decisive factors are at play in the progress of innovation. With a holistic approach, in this paragraph I analyze the direct influence of two non-technological factors, specifically market competition in the supply and regulation with the influence of the indirect impact of two others, namely demand and reflexivity.

Let's first see how the supply ecosystem has changed, precisely how the context of the exchange of goods and services has increased the competition. Platforms[36] are defined around one activity that dominates all others, usually carried out by the payment functionality. In this context, it is possible that BigTech and FinTech

[34] https://www.banque-france.fr/en/financial-stability/financial-stability-mandate/supporting-digital-transformation-financial-sector/wholesale-mnbc.

[35] See the joint Bank of Italy and Bundesbank press release of 4th November 2021.

[36] Platforms are ecosystems in which consumers, vendors, and service providers interact.

companies perform the payment function, replacing the role of financial institutions and taking advantage from data collection on transactions carried out in their networks (Brunnermeier et al., 2021; Segal-Knowles, 2020).[37] Controlling data on transactions would be relevant for pushing in two opposite directions (*ibidem*). On the one hand, digital networks increase competition between digital currencies, reducing the switching costs and creating an un-bundling of the three functions of the currency. Payment platforms, on the contrary, create a re-bundling, associating data-based services and thus creating competition between platforms (and between digital currencies on them). This aspect is relevant, considering the tendency towards the concentration of large players (BigTech) who offer an increasingly large number of services on their platforms. Transactions within these platforms can take place beyond normal national currency borders, identifying "digital currency areas", when the network uses its own unit of account or a payment instrument is valid only within its network (Brunnermeier et al., 2022). In this context, the impacts on the development of digital currencies can be significant. The government response (or lack of intervention) in terms of market concentration will have great importance on the possible evolution of digital currencies on these networks. This competitive environment goes beyond the areas of intervention of the monetary authorities.

Digital networks lead to further competitive pressures, especially on costs of cross-border transactions. This market segment has not experienced particular developments in the past decades, maintaining a significant share of the value chain between intermediation players.[38] Some numbers may be useful to understand the margin available for new solutions: (1) global remittances have an average of 6.25% cost of the amount sent (The World Bank Group, 2023); (2) financial institutions, facilitating a cross-border payment, were estimated to receive on average 20 dollars in fees from a single transaction (McKinsey and Company, 2018); (3) in the euro area, Fintech providers charge total fees between 0.74 and 4.12 euro, while the cheapest bank offers 4.79 euro (Bindseil & Pantelopoulos, 2022). The margins of efficiency for disintermediation are therefore present. Stablecoins, in this context, represent the natural competitive development of the free market. They could reduce inefficiencies and bring direct benefits to consumers. This specific competitive drive also affects the authorities. The latter, driven by the prospect of a crowding out by stablecoins, have taken steps to innovate cross-border payment processes that allow for better interconnection between the different payment systems.[39]

Additional competitive pressure occurs on commercial banks from Bigtech and Fintech companies through the extension of payment and financial services to the overall product and service offerings of their platforms. Banks will need to increase their management and processing skills in communication and information (Lanotte

[37] The authors refer to examples such as Alipay e WeChat Pay in China.

[38] The number of intermediaries could vary from one to five for euro-denominated cross-border payments (BIS, 2021).

[39] See in this regard the work of the BIS on the cross-border payments and cross-border via stablecoin (Brainard, 2021).

& Trapanese, 2023). This is particularly relevant in the European market, where the financing of the economy is more bank-centric compared with the U.S. market-based model. A shift in market weights, particularly the role of banks, will affect as seen the equilibria of the two-tiered system, in a way that is not easily predictable. The ability of the banking market to respond to this challenge, linked to how the regulator chooses to protect it, will be decisive in defining the configuration of the digital currency of the future.

Market competition therefore pushes the supply to develop new technological solutions aimed mainly at disintermediation in the value chain. Indeed, the competitive advantage does not necessarily come from the issuance of a "better currency," but often in reducing intermediary players in the payment services value chain. In this sense, competition is the real driver of innovation of digital money, by way of technology, and is pursued mainly through disintermediation in at least three different use cases.

First, disintermediation leads to the development of DLT technology which implies a reduction of players in the value chain, in particular crypto-assets peer-to-peer transaction would be to disintermediate the banking system over the payments. As represented, the pursuit of the functionalities of a "good digital money" are very complex and require studies and adaptations over time not easily pursued merely by a single technological generation. Technology is never final and requires development and adaptation over time. A single technology, such as Bitcoin's PoW-based blockchain, generally cannot provide the necessary functionality over time or have the speed of adaptation needed to generate trust as a transactional tool for a long time. In that sense, I would expect additional evolutions.

Second, disintermediation drives the development of alternative payment services that aim to shorten the value chain as in the mentioned cross-border transaction market. In this area, the competition generated by private sectors with the asset-backed stablecoins has indeed stimulated the innovative process of monetary authorities (see Auer et al., 2021). These have been active in establishing arrangements and processes for managing international money flows at a lower cost. It is important to highlight also the role of the demand for digital currency, a crucial factor closely correlated with supply. It is precisely the latent demand for transactions at lower costs that has defined the use case for the development of asset-linked stablecoins aimed at cross-border payments.

Third, disintermediation generates new fragmented ecosystems, especially those based on financial services developed on DLT technology. These ecosystems are characterized by a different configuration of the value chain, where technology providers collaborate or compete with traditional financial intermediaries. The Bank of Italy, with a public communication on DLT and crypto-assets, raised the market's attention to the fact that such a fragmentation of responsibility could generate new or additional risks both for the banking institutions that will use them and for the technology suppliers on which these services are based (Bank of Italy, 2022). This kind of innovation is redesigning operators' strategies and business models, amplifying the interconnections in the financial system, mutual dependencies and overcoming national borders (Perrazelli, 2023). The control structure of the supervisory authorities will be a major challenge for the immediate future.

Directly related to the supply there is the demand that, indirectly, strongly influences the initiatives towards innovations. A correct analysis of demand, which is usually more complex and unpredictable than the analysis of supply, can lead to a more precise prediction of market evolution. To understand the future development of digital money, it will be fundamental to use specific tools of analysis, as those of marketing, used by the private sector in search for business growth. In this context I limit myself to highlighting the relevance of demand as a factor of innovation, while will be necessary a specific analysis on the marketing approach needed for user needs study.

Regulation also plays a significant role in the development of innovation. Money can be considered a social institution, not a mere technological support, for which there has always been a history of interconnection between progress and institutional adaptations (De Bonis & Ferrero, 2022a, 2022b). In this perspective, technological innovation refers to digitization which is changing various sectors of society, especially finance. The adaptation, which we are still experiencing, will therefore also provide for a rethinking of the role and tasks of the institutions, already widely underway in the European Union.[40]

Regulation has multiple effects. It shows the way to pursue good innovation, creating appropriate limits for socially undesirable services, but also offers guarantees for the development of certain financial services. Regulation is in fact a tool to direct the behavior of companies towards safer and more sustainable solutions and to protect the more responsible operators from the unfair competition of the less virtuous ones (Perrazelli, 2021). Clearly, being a sector with a high rate of innovation, regulatory interventions will follow also with updates over time. The relationship between rules and innovation is in fact characterized by an asymmetrical speed of development, an innate slowness on one side and a frenetic speed on the other. The guiding function of regulation is one of the most discussed but crucial aspects of innovation.

On the other hand, however, regulation can also generate reflexive phenomena. As mentioned, regulation shapes the reference markets and consequently generates adaptation reactions from enterprises. Some of these are foreseen or desired, while others are unexpected or undesired. Regulators should therefore also consider the potential reflexive effects on the market (Ippoliti, 2020). These effects are not always easily predictable and could have unexpected consequences that are potentially even contrary to the objectives of the regulation.[41] The possible reflexive effects

[40] Consider, for example, the aforementioned MICAR, the Digital Finance Package, the PISA framework, the DLT Pilot Regime.

[41] For example, Basel regulation had foreseen a risk-weighted rating in which Eurozone debt securities had the same weighting. The consequence was an underestimation of Greek debt with consequences for bank balance sheets (Ippoliti, 2020).

of regulation are a factor to be taken into serious consideration in the development of new forms of money.

A theoretical example of possible reflexive effect could come from the application of the emerging regulation, where different rules are applied to different types of crypto-assets. This could lead to the development of solutions in the market for types with less regulatory burden. As an extreme case, it could lead to a greater concentration of services with crypto-assets class currently not covered, which could temporary allow the development of initiatives outside the regulatory objectives. For example, this scenario could occur for the NFTs, currently not covered by MiCAR but defined in its initial recitals, as to avoid extensive interpretations of crypto-assets to fill in this category.[42] While this is an extreme possibility of reflexivity, more clearly one can imagine how different burdens on specific crypto-assets categories could drive developments for solutions potentially away from the goal of the regulations.

Another possibility of a reflective effect could come from the co-presence of multiple specific regulations in the same geographical area. For example, In the context of crypto-assets, different regulations can jointly apply: payments, investments, AML, etc. The coherence between different regulations is neither obvious nor trivial.

A different perspective on the reflective effects of regulation concerns the comparison between different regulatory perimeters, deriving from the "regulatory attraction". A crypto-asset can in fact be defined depending on the relevant legislation. For example, the widespread market denomination of the crypto-assets "stablecoins" differs greatly from the terminology used in the European MiCAR, where indeed the difference from the use of market terminological reference "stable" recalls a conceptual and substantial distance with respect to the actual monetary characteristic of stability. The crypto-assets can fall into three different categories in MiCAR depending on their technical characteristics: electronic money token, asset reference token as well as crypto-asset other than.[43] Other categories are defined but out of MiCAR scope, such as NFTs and security tokens.

This perspective evidently looks at the more ontological aspects of crypto-assets and highlights how their future development and diffusion can actually reflect the different correspondence between regulations. A future evolutionary dynamic will also concern these dynamics. For example, European regulation, with its pioneering intervention on the subject, could have effects also on the development of crypto-assets initiatives by those companies located outside the regional perimeter of competence but characterized by projects with an international perspective (e.g.

[42] MiCAR, whereas (11): "The fractional parts of a unique and non-fungible crypto-asset should not be considered unique and non-fungible. The issuance of crypto-assets as non-fungible tokens in a large series or collection should be considered an indicator of their fungibility. The mere attribution of a unique identifier to a crypto-asset is not, in and of itself, sufficient to classify it as unique and non-fungible. [. . .]".

[43] The so called "algorithmic stablecoin", with lack of collateral assets in the stabilization mechanism, would fall in the "crypto-assets other than".

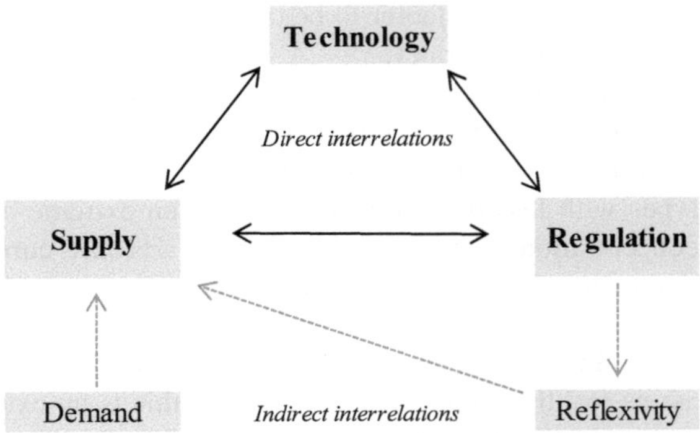

Fig. 11.3 Direct and indirect interrelations with technology

global stablecoins). The indirect reflective effects on innovation also deserve a dedicated in-depth study.

Summing up, when talking about technological innovation, as is the case with multiple studies on the subject, we should consider also the direct interrelations of the technology with the supply and with the regulation and, moreover, the indirect effect of the demand and reflexivity (Fig. 11.3).

The main driving forces behind innovation are at least three: technology, supply and regulation. These forces are directly interconnected, the development of one significantly alters the equilibrium condition of the others. In this historical moment, the three forces are moving simultaneously, therefore making it particularly difficult to predict developments of its market in the medium to long term.

Supply, with the competitive pressure on research and development, is relevant on the creation of new technological solutions. Competition takes place especially in forms of disintermediation, pursuing the goal of reducing subjects active in the value chain. In addition, the definition of new products and services highlights the need for new regulatory interventions. The evolution from the first blockchain and the consequent diffusion of thousands of crypto-asset, as well as the need for a new regulatory perimeter, are an example of the impact of supply respectively on technology and regulation.

The emerging regulation directly influences the market supply, with a function of directing the desired development, both defining competitive perimeter and setting limits on the use of technological innovation. These effects will be particularly relevant within the European Union, where the relevant regulation has moved in advance with respect to the international context.

In addition, other relevant factors indirectly influence the three main forces: the demand and the reflexive effects should be taken into consideration in a holistic approach. These two are probably underestimated and they could have in the future a relevant impact on the development of the new form of digital money.

The reflective effects, consequence of the reaction and adaptation of market supply to the regulation, generate specific new product and service development. In this sense, regulation, through the indirect means of reflexivity, further influences specific market innovations. The theoretical example presented above, the potential influence of the future implementation of MiCAR on the development of specific crypto-asset solutions, shows that reflexivity is an indirect aspect to be taken into due consideration. In addition, the cited case of the Paypal stablecoin, in line with the emerging MiCAR regulation, could be an example of reflexive effect on the supply and on the use of technology for a company outside the regional regulatory perimeter.

Demand, latent or emerging, directly drives the direction of supply. Demand, indirectly through the initiative of offering goods and services, is indeed a factor in the development of technological innovation. The mentioned case of spread of crypto-assets for cross-border payments use case could be an example of demand-driven innovation seeking market shares by satisfying user needs.

11.6 Conclusion: The Future of Digital Money

We have initially highlighted four elements of money: functionality, trust, nature (public/private) and creation. Using these fundamental concepts, we evaluated the market from the supply side prospective. Technology represents an important innovation driver that can open up new opportunities in the market, but it is not the only one. In the supply, the competitive factor of digital networks has also played a fundamental role in the innovation of money market. Competitive forces have led to the search for greater economic efficiency, led by the demand and realized with disintermediation in the value chain. Finally, regulation, by defining the desired development perimeter but also potentially producing reflexive effects, intervenes in development and conditions of innovation.

The future of digital money cannot ignore the availability of the three functions envisaged until now. Consequently, it is difficult to imagine that the payment system, the control of the unit value of money over time and the supervision of credit creation can be carried out by a non-public entity. In the medium-short term, I would therefore exclude the emergence of a model other than the existing two-tiered system.

New instruments are now being added to those already available. The emerging regulation provides for the possibility of using some type of stablecoins with characteristics and functions similar to those of electronic money. Market operators will define new uses and new services, for which the demand for this new type of digital currency will be destined.

It is more difficult to predict what the reflexive effects of regulation will be. Also in this case, the entrepreneurial activity will find solutions compatible with the legislation but that maximize profits. The alignment between the objectives of regulation and those of entrepreneurial activity is far from resolved. The high competition in the market will most likely lead to find new applications of existing technology or to develop new technologies. Innovation could drive for opportunities on reducing the local regulatory burning (e.g. companies in EU) or, as opposite, following the direction set by regulation of other countries (e.g. companies in USA).

This intrinsic unpredictability certainly makes it difficult to study the reflexive effects of regulation, but it makes the need for further study even more current and concrete.

If the monetary anchor between public and private money appears to be a clear element on the desirability of a CBDC, the complexity of the monetary system makes it difficult to predict how and when it will eventually be introduced. The introduction of a CBDC into the system, or its non-issuance, could alter the current mechanisms in a more or less predictable way. The capacity and competence of the monetary authorities must in fact deal with the unpredictability of a changing demand for money for different purposes. Until now, Central banks have operated under a monopoly regime for the money supply, both for retail banknotes[44] and digital currency for wholesale payments.

In the event of a retail CBDC issue, the big challenge is the possibility of issuing a currency in a market with other retail digital currencies (private and public), potentially altering the balance of the monetary anchor. In fact, if the retail CBDC becomes too widespread, it could crowd out the money of banking intermediaries, putting the mechanism of maturities transformation and credit creation at risk. Instead, if CBDC is not successful, the objectives pursued could have not be achieved. This last scenario would be equivalent to the decision of not issuing a CBDC. However, it would still expose the public money to other threats, such as that of reducing the monetary anchor in case of strong reduction of banknotes or of a dollarization by another foreign retail CBDC.

This new competitive scenario for the Central bank will be the most critical aspect of the future of digital currencies. As highlighted, the competitive aspects are many and reflect the research of demand for products and services. In any case, the authorities should take steps to develop new skills, especially in the analysis of market demand, to predict with greater precision what are the technical-functional characteristics that will be required for digital transactions. Therefore, it is no longer just a matter of estimating the quantitative demand of money (as is already the case), but also to make analysis on the present or latent demand for specific functional characteristics of digital money. This would require a significant change of approach to be taken into account in the current organizational arrangements.

Finally, one last consideration on the availability of non-digital currencies. The European Central Bank has clearly represented its intention, in case of a retail CDBC issuance, not to cease supplying the currency in the form of banknotes. This shareable position is expressed according to the objectives of financial inclusion, thinking for example of the need of avoiding exclusion of generations not present in the digital network. From this aspect comes a consideration on the resilience of the system of a cash-less society. The scenario in which the backbone infrastructures of the digital network fails would expose the entire payment system to a dangerous unavailability. In an analogy with the distribution of banknotes via ATMs throughout

[44] In reality, some central banks have already experienced a sort of competition on the banknotes of weak currencies, such as the case of the argentine peso with the dollar.

the national territory, we should perhaps consider a more distributed supply of electricity as an essential element for the business continuity for digital currencies in the region. Thinking in terms of resilience is useful and necessary in order to prepare economic systems to natural threats or potential attacks from hostile countries. The latter case, even in merely manifesting itself as a possible danger, would generate consequences and represent a point of vulnerability for society as a whole. Lasting digital currencies will also depend on the resilience of the entire system.

References

Ardizzi, G., Nobili, A., & Rocco, G. (2020). *A game changer in payment habits: Evidence from daily data during a pandemic* (Bank of Italy Occasional Paper No. 591).

Auer, R., & Bohme, R. (2020). *The technology of retail central bank digital currency.* BIS quarterly review.

Auer, R., Haene, P., & Holden, H. (2021) *Multi-CBDC arrangements and the future of crossborder payment* (BIS papers No 115).

Auer, R., Haslhofer, B., Kitzler, S., Saggese, P., & Victor, F. (2023). *The technology of decentralized finance* (DeFi) (BIS Working Papers No 1066).

Bank of Italy. (2021). *Banca d'Italia and Deutsche Bundesbank discuss initiatives for DLT-based asset settlement in central bank money*. Press release.

Bank of Italy. (2022). *Communication by Banca d'Italia on decentralized technology in finance and crypto-assets*.

Bank of Italy. (2023). *Banknote migration and the estimation of circulation in euro area countries: The Italian case* (No. 40).

Bilotta, N., & Botti, F. (2021). *The (near future) of central bank digital currency*. Peter Lang.

Bindseil, U. (2019). *Central bank digital currency - Financial system implications and control.*

Bindseil, U. (2020). *Tiered CBDC and the financial system* (ECB Working Papers Series).

Bindseil, U., & Pantelopoulos, G. (2022, August). *Towards the holy grail of cross-border payments* (ECB Working Paper Series No 2693).

BIS. (2019). *Investigating the impact of global stablecoins.* G7 Working Group on Stablecoins.

BIS. (2021). *Central bank digital currencies for cross-border payments.* Report to the G20.

BIS. (2023). *The crypto ecosystem: Key elements and risks.* Report submitted to the G20 Finance Ministers and Central Bank Governors.

Brainard, L. (2021). *Private money and central bank money as payments go digital: An update on CBDCs.*

Brunnermeier, M. K., James, H., & Landau, J. P. (2021). *The digitalization of money* (BIS Working Papers No 941).

Brunnermeier, M. K., James, H., & Landau, J. P. (2022). *The digital euro: Policy implications and perspectives*. Policy Department for Economic, at the request of the committee on Economic and Monetary Affairs (ECON).

Brunton, F. (2019). *Digital cash. The unknown history of the anarchist, utopians and technologists who built cryptocurrency*. Princeton University Press.

Bullmann, D., Klemm, J., & Pinna, A. (2019). *In search for stability in crypto-assets: Are stablecoins the solution?* (Occasional Paper ECB).

De Bonis, R., & Ferrero, G. (2022a). *La moneta digitale di banca centrale: progresso tecnologico ed evoluzione delle istituzioni* (Banca d'Italia Occasional Papers N. 690).

De Bonis, R., & Ferrero, G. (2022b). *Tecnologia, finanza, moneta e istituzioni. Confindustria servizi.* Rivista di Politica Economica.

European Central Bank. (2020, October 2). *Report on digital Euro.* ECB website.

Ferrari, A., & Ferrero, G. (2020). *Moneta pubblica e privata in un sistema economico moderno*, mimeo.
Guala, F. (2016). *Understanding institutions: The science and philosophy of living together*. Princeton University Press.
IMF. (2012, September). *What is money?* Finance and Development.
Ippoliti, E. (2020). *Un filosofo a Wall Street. Speculazioni sulla finanza da Aristotele ai bitcoin*. Egea.
Lanotte, M., & Trapanese, M. (2023). *Financial intermediation and new technology: Theoretical and regulatory implications from digital financial markets* (Bank of Italy Occasional Papers N. 758).
Maimeri, F., & Mancini, M. (2019). *Financial intermediation and new technology: Theoretical and regulatory implications from digital financial markets* (Bank of Italy Occasional paper No. 758).
McKinsey and Company. (2018). *A vision for the future of cross-border payments*. Global Banking practice.
Panetta, F. (2022). *Crypto dominos: The bursting crypto bubbles and the destiny of digital finance*. Keynote speech at the Insight Summit held at the London Business School.
Panetta, F. (2023, November). *The macroeconomic implications of central bank digital currencies*. Joint CEPR - ECB Conference, keynote speech.
Perrazelli, A. (2021). *FinTech regulatory initiatives: Where are we now?* Speech at Università degli Studi dell'Insubria Laboratorio di Finanza Digitale.
Perrazelli, A. (2023). *Foreign banks in Italy: Regulation and innovation*. Speech at the meeting of the General Council of the Association of Foreign Banks in Italy (AIBE).
Segal-Knowles, C. (2020). *Responding to leaps in payments: From unbundling to stablecoins*. Bank of England, speech.
Sveriges Riksbank. (2017, September). *The Riksbank's e-krona project Report 1*.
The World Bank Group. (2023, March). *Remittance prices worldwide quarterly*, issue 45.

MIX
Papier aus verantwortungsvollen Quellen
Paper from responsible sources
FSC® C105338

FSC
www.fsc.org

If you have any concerns about our products,
you can contact us on
ProductSafety@springernature.com

In case Publisher is established outside the EU,
the EU authorized representative is:
Springer Nature Customer Service Center GmbH
Europaplatz 3, 69115 Heidelberg, Germany

Printed by Libri Plureos GmbH
in Hamburg, Germany